# Space X

## Starlink and Our Man Elon Musk

*Bringing NASA Astronauts to the space station
and Mega Project To Colonize mars*

## Michael Barnes

Published By **Oliver Leish**

# Michael Barnes

*Space X: Starlink and Our Man Elon Musk Bringing NASA Astronauts to the space station and Mega Project To Colonize mars*

**ISBN   978-1-77485-884-4**

Legal & Disclaimer

The information contained in this ebook is not designed to replace or take the place of any form of medicine or professional medical advice. The information in this ebook has been provided for educational & entertainment purposes only.

The information contained in this book has been compiled from sources deemed reliable, and it is accurate to the best of the Author's knowledge; however, the Author cannot guarantee its accuracy and validity and cannot be held liable for any errors or omissions. Changes are periodically made to this book. You must consult your doctor or get professional medical advice before using any of the suggested remedies, techniques, or information in this book.

Upon using the information contained in this book, you agree to hold harmless the Author from and against any damages, costs, and expenses, including any legal fees potentially resulting from the application of any of the information provided by this guide. This disclaimer applies to any damages or injury caused by the use and application, whether directly or indirectly, of any advice or information presented, whether for breach of contract, tort, negligence, personal injury, criminal intent, or under any other cause of action.

You agree to accept all risks of using the information presented inside this book. You need to consult a professional medical practitioner in order to ensure you are both able and healthy enough to participate in this program.

## TABLE OF CONTENTS

## Introduction

If you're looking to understand what Elon Must is the man behind SpaceX, a trip to the SpaceX headquarters located in Hawthorne, California, will reveal a lot about his personality. Visitors who step inside the space are greeted by two huge posters that showcase Mars in the full glory. They are the entrance to the Musk's cubicle.

Why are these posters so significant?

The poster to the left depicts Mars as a deserted land that's nothing more than an orange orb. the one on the right depicts an earth covered in green, that is surrounded by oceans. Mars is hot and has been changed in order to allow humans to reside on the planet. Musk has a vision. Musk is pursuing, and he is determined to realize it.

If we could create a sustainable method of energy sources, we could become a multi-planetary species that could be able to survive on another planet, and withstand the worst-case scenario which could endanger humanity's existence.

On the 28th of June, Elon Musk hit another milestone: he turned 50. In the past 50 years, Musk has accomplished the impossible. Musk became the Chief Executive Officer of SpaceX and Tesla and founded Neuralink and OpenAI The Boring Company, and co-founded Neuralink. The Boring Company, and was determined to fulfill his long-standing dream to leave Earth for Mars.

If you take a moment to think about it, solar-powered batteries, electric vehicles as well as space rockets and the billions of dollars in wealth create Musk an actual Tony Stark. If you're not aware that the source of inspiration for"Iron Man," the year 2008's Marvel movie "Iron Man" is Elon Musk.

In contrast to other billionaires who boast the wealth of their families, Musk has a greater focus on coming up with new ways to create a better place. The year was January 20, 2021. Tesla Owners of Silicon Valley tweeted "@elonmusk is now the wealthiest person on earth with $90 billion."

Tesla's CEO Tesla replied to this by tweeting this tweet

"How bizarre" and "Well it's back to work." ..."

This tweet offers a fascinating information about what Musk believes about his fortune. Many refer to SpaceX's SpaceX office "Muskland," and it makes visitors awestruck with their mouths open as they walk into the building. The front facade is covered in a 550,000-square-foot rectangular which reads "Unity of Mind, Body, and the Mind" with white letters.

Although the building may be in a poor neighborhood however, you are greeted with its majesty when the doors of SpaceX's SpaceX building are opened. SpaceX offers a number of structures where rockets are constructed at the same time. Tesla owners can go to the parking area outside the SpaceX studio, where cars can be charged with electricity at no cost.

However it is true that the Muskian character of Elon isn't easy to accept because he loves to be in control of the world around him. In 2015 an American business journalist Ashlee Vance was enthralled to tell the untold tale about Elon Musk. After advising Musk about his intentions Vance was told with a stern tone that the billionaire would not be a good partner for Musk. After interviewing a number of Musk's close friends as well as former and current employee, Vance still hadn't learned anything new that could not be located online. In the course of 18 years talks, Musk finally agreed to meet with Vance. He was willing to cooperate with Vance for as long as Musk had the opportunity to review the novel prior to when it was published. Vance was aware of this, however, for a variety of reasons, did not comply. He was of the opinion that Musk's story could be his view rather than the best version of everyone else and would not do justice to the facts. But, he was willing with Musk and to take his own side of the story into consideration.

## Chapter 1: The Myth Of Elon Musk

"When something is significant enough, you will do it even when the odds aren't to your advantage."

Life is unpredictable. Have you been told this?

There is a reason for everything however, you can alter your own story however you'd like to be the cause. All that's needed is determination and dedication. Failure is, was always an possibility, so don't be worried about it. If you fail, you'll learn by doing what they say "Try and keep trying until you're unsuccessful."

This is the way Elon Musk became an businessman of the century, and gradually climbed his climb to the highest levels!

Musk was born on the year 1971 on the 28th of June located in Pretoria, South Africa. As with all kids, Musk had big dreams that were centered around learning something that was new. His tendency to daydream got his parents concerned which led the doctor to recommend an hearing test.

We now know Musk's sole handicap was his aversion to advice that did not encourage him.

When Musk was 10years old, He began to show a love for the field of computer science and technology. This development in him was because of his parents' divorced. It was 1984 when the public first became exposed to Elon Musk. He was only 12 when he first met him. PC and Office Technology, an South African trade publication, discovered the source code for the video game Musk had created. The space-based science fiction game "Blastar" required about 167 lines of commands in order to play at the point that a command had to be entered for the computer to complete something. Although his game didn't perform as well as he had imagined, it was certainly better than the ideas 12 year olds had come up with. The game Musk was covered in an Magazine and his contribution was worth 500 dollars. The spread for the game that was published in the 69th page, depicted the young man E. R. Musk and his dreams of conquering.

The game's description read "You must destroy an alien space freighter that is carrying dangerous Hydrogen Explosives and Status Beam Machines." At the age of just a few years, Musk was a remarkable and brilliant student. His ability to combine both reality and imagination led him to come up with the ingenious ideas that result billions of dollars in the present.

While he was an inventor at 12 , and appeared in a variety of magazines, people was not willing to accept his work. After releasing his first software everywhere but he still was a long way to take. Musk understood how important education is and devoted the majority of his time learning. Being a bibliophile and an introvert Musk was bullied at grade school.

At the age of 14, Musk went through an existence crisis. Being an adolescent, but an exceptionally gifted one Musk dealt with the issue as most adolescents do by turning to religious and philosophical texts. He tried several different ideas, but ended up back in the same place he began and then embraced science-fiction lessons. Through this he discovered that the sole way that a individual could consider thinking out of the box, or get out of their comfortable zone was to strive to attain greater insight.

At the age of 15, Musk went through an increase in size that enabled him to stand up for himself. Although Musk wasn't the type of person who enjoyed combat or fighting however, he did learn to grapple and Karate.

At the age of 17 Musk relocated to Canada. The year 1989 was the first time he was enrolled in Queen's University and avoided the South African military service that was required at reaching a certain age. At the time, Musk was able to obtain an Canadian citizenship. Musk believed that, having a Canadian citizenship it would be easy to obtain another within the US.

The year 1992 was the time that Musk moved to US to learn about physics and commerce with the University of

Pennsylvania. After earning his bachelor's degree focused on economics, Musk returned to pursue a different degree in the field of physics. He was always hungry for knowledge, which is why the Californian enrolled him at Stanford University for a Ph.D. in energy Physics. In the year Musk's fortune took off. The boom in the internet era took over and Musk left Stanford to create his first venture.

In the end, Zip2 Corporation was born in 1995.

Zip2's Story of Zip2 began in 1994, when Musk was working on his internships. While inside the workplace, he engaged in an exchange with the Yellow Pages' salesperson. Musk tried to sell his idea of creating an online listing in addition to the paper version. The salesperson was not equipped with any understanding of the internet and how people could use it to locate businesses and so, Musk struggled to comprehend the idea.

Though Musk's pitch was not convincing but he was convinced and decided to relay the idea the story to his older brother Kimbal. Because Kimbal himself was in the field and recognized the potential of the concept and the two brothers determined to take action to address the issue. The brothers then formed Global Link Information Network, which later became known as Zip2.

Many companies did not know how the internet could help the company. This is why they didn't have any intention of having a site developed that could allow them to market online. This was a concept that was not yet born.

Thus, Musk and Kimbal hoped that businesses like clothing stores as well as hairdressers and restaurant owners would appreciate the benefits from it.

The concept behind Zip2 was to make a business directory which was linked to maps. It would allow people to look up businesses, and provide instructions to them. It's possible to dismiss this idea, but we are able to use Google Maps and Yelp, however, no one was able to create such a service in the past.

Zip2 was created in the hands of the Musk brothers from Palo Alto at 430 Sherman Avenue. Because the Musk brothers were lacking money in the early days they took a room to house their office. It was 20 feet x 30 feet and bought a few pieces of furniture. The 3-story structure where the business was located was without elevators.

Every time it would happen that the toilets would fill up. According Musk, according to Musk the place was an "shitty spot" to begin a business and that's what they could manage in the moment.

One of the things that was essential for the startup was a speedy internet connection. Thus, Musk asked entrepreneur Ray Girouard who was also an Internet service supplier and who had a desk located on the lower floor give him an internet connection.

Zip2's code was managed by Musk and Kimbal, while Kimbal was in charge of door-to-door sales. Once everything was in place, Musk contacted Navteq, an organization that developed digital maps, and made an agreement with them. Zip2 and Navteq joined their databases, and developed an insufficient system. The company started with just $28,000. By when the map was completed and the maps were published, both Musk as well as Kimbal were broke, so they were living in the office. They shopped on the YMCA and ate four meals per day at Jack at the Box.

As time passed, Musk made more improvements to the code, and eventually, they began hiring. However, they were short of capital , but an opportunity was to his doorstep. The director of a venture capital company known as Mohr Davidow Ventures was aware of the fact that two South African boys were making Yellow Pages for the internet. Then, he invited them to visit with a pitch. He was so impressed by the concept they presented that he decided to invest $3 million in the business.

Zip2 later moved into a larger office and developed a product specifically aimed at newspapers. The print media took a long time to grasp how the internet could help them, but eventually they recognized the advantages and hopped onto Zip2's Zip2 train.

The investment forced Musk to the position as Chief Technology Officer as well as Rich Sorkin was hired as the company's CEO. Even to this day Musk is adamant about the moment when he made the decision to let go of the Zip2 company and appointed Sorkin his CEO. After a time Musk felt the deal was with the evil one.

The problems began when the new employees began making changes to the code. Because Musk had been a self-taught programmer and his abilities were not a comparable to what other engineers were able to offer. He was not a fan of the coding modifications but they were needed because Musk's work was too complicatedand, because of it, the program could be prone to crash.

As the company continued reach new highs, Musk felt that Zip2 was heading in a totally different direction. The company was now focused on catering to businesses instead of consumers. Thus, Musk decided to pitch the idea of buying an entirely new domain name to sell their services to customers. The problem was that Sorkin was enthralled by the publicity and revenue they received by media corporations. He chose to merge with CitySearch.com that provided a few of similar services to Zip2. The announcement of the merger appeared in the local newspaper and shortly thereafter, the two companies had to make the tough decision of which employees to retain and which ones to let go. Because CitySearch.com had a stronger group of engineers Zip2 employees knew that they were going to be fired first. Musk was the first to endorse the merger, subsequently backed down the merger.

What began in 1995 was over in 1998 after the merger was scrapped. Musk called a board session and asked the board members to eliminate Sorkin and reinstate Sorkin as SEO. But the board refused and Musk was dismissed from the chairmanship. In the following days, Sorkin was replaced by Derek Proudian.

Musk requested Proudian Musk if he could support the idea of a service that is based on consumer demand however, Proudian was not convinced and stated, "This is your first company, so is the time to acquire an acquisition partner and get started on your third, second, and fourth business."

Microsoft was already present into the market, and was heading along the same path. Proudian was of the opinion that investing additional money in the company would result in losses. As of 1999, Compaq Computer made an offer to purchase Zip2 at a price of $307million. Everyone was in agreement it was the most effective method to overcome any loss they'd suffered in the past.

Musk along with Kimbal received $20 million , and $15 million in exchange for their shares in the company and they haven't looked back. After Zip2 being sold, Musk moved on to his next venture, and, from then on Musk made a decision to not relinquish his rights to being the CEO of any of his businesses in the future.

The most significant lesson Musk learned during his very first business venture was that the way he handled things could have been slightly differently. Because he'd never worked in a group prior to this, he had no idea of how to describe his idea to his coworkers. The employees would leave in the evening, but Musk would remain behind working to make the code better. They would return to work the next day and find the work they had done altered. Furthermore, Musk was never good in apologizing and his abrasive style of speech never went down well with anyone at the company.

## Chapter 2: The Background Of Paypal

It all began in 1999. Musk was ebullient following the sale of Zip2. He had learned many things in the background and had achieved his goal of becoming a millionaire on the dot-com. Then, he was looking for a sector that was lagging behind and was prone to inefficiencies, but also had plenty of cash. He was keen to explore the internet more thoroughly, and after some thought and research, he came across the perfect area to target banks.

In his time as an intern been employed at The Bank of Nova Scotia. While there, he learned that even though bankers were wealthy however, they were also apathetic and this was a huge opportunity that was worth investigating. At the time when he was Director of Strategy, and was had been asked to review the portfolio of debt from third world countries of the firm. The debt in this portfolio was classified as "less-developed countries' credit." Countries on South America and other continents have been in default several times and made it necessary for the bank to not take on the growing debt. Musk was advised to dig through the bank's holdings to discover the amount of the country was in debt.

That's where Musk discovered the opportunity to start a business and took the decision to address the massive debts were owed by the US had. To ease the burden of debt it was the US created Brady Bonds that backstopped the obligations of other countries like Argentina or Brazil.

The arbitrage strategy wasn't without being noticed by Musk. The backstop price was 50 cents, while the actual debt was trading the 25-cent range.

Musk was jumping in excitement to take advantage of this chance by his feet, hands and even his brain. The entrepreneur casually sought out information about Brazilian debt that was owed to an exchanger on the market. When asked what amount he would like for 25 cents, the trader put his first thought that came into his head $10 billion.

The trader stated that it was possible then Musk called off the meeting. He was stunned by the comments of the trader since the money was provided with Uncle Sam. The previous summer, he been paid $14/hour and exasperated for using the machine which was reserved for the top executives. After the conversation he had, the chance for him to make a mark was his face. He ran into his boss and told him "You could have your debt settled at no cost."

The boss ordered him to prepare the report. It was later passed on to the CEO who turned it down, claiming they'd been burned often in the past by Argentina and Brazil and were not willing to take further chances.

Musk described to the company how they had been funded by Uncle Sam however the CEO would not take his advice. Following the rejection, Musk considered opening his own online bank and discussing the idea with his colleagues from Pinnacle Research. In his efforts to convince researchers of his plans however, nobody was as invested in the same way Musk was because they all believed that it would take years to make sure that the security of internet is top-quality enough so that customers could be able to trust it.

The scheme Musk had devised was way unimaginable to anyone's eyes. There was a feeling of skepticism about purchasing books on the internet, and here, Musk was planning to explain to people that by entering the account numbers into their accounts they could take advantage of the advantages of banking online.

The idea was to create an institution for financial transactions with savings and checking accounts, along with brokerage and insurance services. At the time, constructing such a system was not a problem. But, the regulations and rules for making an online publication, was also made from scratch, were previously unheard of. However, Musk always the optimist was confident that he could accomplish it. This wasn't a venture similar to Zip2 in which all Musk needed to do was to provide direction. Musk was asking clients to trust Musk with their financials and, if something were to happen to the services offered the consequences could be catastrophic. Despite the numerous negatives to Musk's plan, Musk jumped into action.

Prior to the time that the purchase of Zip2 was finalized, Musk had started working on his new strategy. He spoke to the top engineers that had worked in Zip2 and even got in touch with his former colleagues at some bankers in Canada.

In March 1999, Musk incorporated X.com. Since it was a finance-focused start-up the name was somewhat sexually provocative, however Musk did not seem to be concerned about it. At this point, he had gained a reputation as a billionaire and was enjoying the extravagant lifestyle he enjoyed. It was reported that he made a call to CNN in the early hours of seven A.M. to witness the arrival of his McLaren model, which was the only one of there were 62 McLarens around the globe.

With the $22 million the man had become confident and his flashy lines before the cameraman caused the crew to shiver. He spoke about the Zip2 deal, and then took the viewers from their homes into his X.com office.

Despite his snarky behavior like blasting up his McLaren and speaking too much about money, and flinging the wealth of his family around Musk was not the look of a playboy. The bulk of his wealth in fact, over half was spent on the development of X.com. At the end of the day the company was sold to him for the sum of $4 million for personal use.

Musk put about $12 million in X.com and left him after taxation, with around approximately $4 million to use for personal purposes. "That's one of the things that separates Elon from other people," said Ed Ho who was the former Zip2 executive who was later the co-founder of X.com. "He's prepared to accept on an incredible amount of risk for himself. When you sign a contract such as that and it pays off, or you get stuck in a bus stop or someplace."

Musk's choice to invest entirely in was a massive error that could have ended in Musk living in his car. But, he was willing to risk his life, much more that his shareholders. Zip2 was a clever and simple idea, but X.com could change the world.

The thing people learned about Musk who became his signature was the fact that he embarked on a business that was extremely complex, with a lot of planning, but no understanding of the specifics of the business. The way that Musk accepted this proved that he did not care in about anything in the least.

Musk thought that the bankers did not understand the complexities of finance, and Musk had the ability to handle it better than the bankers. His confidence and ego were such that some were enthralled by his work, while some viewed Musk as pompous and unscrupulous.

Then, it was set to go down as the launch of X.com will reveal the flaws of Musk as a trailblazer, unstoppable determination, and awe-inspiring. As we all are aware, with big vision, comes challenges and disappointments that leave leaders feeling empty. Musk didn't know this that he would once more his world is confronted with the devastating reality of being ignored and being a victim of others who are in charge of his business.

Musk began assembling his team that included individuals who were from Zip2 and two finance-related experts Canadians. The group of four was co-founders of X.com and Musk being the most significant shareholder due to his massive initial investment. The cofounders all joined at the beginning, agreeing with Musk's assertion that banks were not up to speed in providing customers with the services they required. In all of them the cofounders, only Musk was aware of the basics of banking. The more that the co-founders learned about the workings of banks and operated, the more they feared about the regulatory concerns that could hinder their establishment.

There were many personalities conflicts. Fricker was who was one of co-founders was determined to leave his mark in the world as banker of the future. The bold statements Musk made didn't go over nicely with Fricker. He viewed them as ridiculous since the company was trying to figure out what it was that they could do. In the midst, Musk and Fricker entered into a heated argument that became a major battle. Fricker was fed up with Musk's erratic thinking and controlling behaviour, which is why he decided to begin an assault.

He informed Musk that he had to allowed him to take over as the CEO or let the employees go and create a business that was his own. Musk has never been successful using blackmail, so he attempted to convince employees to remain with his company. The words he spoke of were ignored and a few of the most important engineers decided to side with Fricker. At the end of the day, Musk was left with only a few dedicated employees and a business which was nothing more than an empty shell of the company it was before.

Musk's mood didn't drop. He just decided to hire more employees. He contacted a few additional investors, among them was Musk's personal friend. The friend decided to back Musk, even though the company was failing. Musk was on a different course to take on the world. He took to the streets with his talk on internet banking in order to draw the most engineers possible. The strategy worked, and hundreds of engineering professionals joined with the business in order to make Musk's dream into reality. After a few months the company was able to secure an FDIC mutual fund license as well as a banking license and enter into an alliance with Barclays. In 1998, the company succeeded in creating one of the very first banks online with FDIC insurance. It secured bank accounts and various mutual funds that customers could pick from.

In the initial testing phase, Musk gave his employees $100,000. Then, on the Thanksgiving holiday, X.com finally went live. The company was involved in a dispute with another company named Confinity which was providing similar services. After Confinity was unable to replenish their cash reserves, Musk decided to form an alliance, making Musk the company's largest shareholder.

As X.com's customers grew and the site's customer base grew, it became difficult for the site to cope with ever-growing demands. Each now and then the website would fail and the constant concurrence from a variety of startups put Musk lots of anxiety.

What Musk did not know was that, once more employees from his firm were planning the perfect coup! A handful employees met in a meeting to discuss ways they could force Musk to leave the firm. The employees then presented their ideas to the board and proposed that Thiel take over the role of Musk. In the meantime, Musk had married and planned to take an excursion to honeymoon along with his spouse. When he stepped off the plane, he received letters circulated regarding the board meeting and those who believed in Musk were shocked to learn that this happened. One of his acquaintances attempted to contact Musk however Musk was not able to be reached. When Musk's plane returned to the United States after which he learned that he was replaced by a new CEO.

As the CEO Thiel changed the name of the company to X.com in 2001 to PayPal in 2001. Musk served as an advisor and invested his money to help the company rise up to the top. The majority of people believed that Musk would be bitter as a result of the change, however, he was surprisingly positive. When PayPal began to grow large-scale companies approached with offers of a buyout. In the early days, PayPal made $240 million a year. So, Musk convinced the board to accept an even bigger deal. Then, in 2002 eBay made an offer of $1.5 billion in exchange for PayPal The board agreed to the offer.

Musk's direction helped PayPal to endure for so long, even when the dotcom boom came to an end. It became an international blockbuster IPO following the terrorist terror attacks of 9/11. After that, it was sold to eBay at such a staggering cost, PayPal left a mark that nobody can erase even in the present.

## Chapter 3: The Story Of Spacex

It was 2002 when the first PayPal transactions took place. Musk was tired of PayPal. After being fired as CEO, he realized that there was no more within Silicon Valley for him. He made the decision to relocate somewhere that was more glamorous and glamorous. The city of Los Angeles that he found the space he was looking for. He believed that he had left internet-related services. In the midst of his childhood fantasies about rocket ships returning to him He changed his course again.

In the evening, as Musk and his team celebrated the purchase of PayPal in Las Vegas, they noticed something new in his personality. Musk was talking openly about Space travel, and the ways it could revolutionize the world. The reason Musk chose LA was the fact that it was a city with regular, mild, and consistent weather and was one of the top cities in Aeronauticals. Musk did not have a strategy to think of however, he knew that in LA it would be surrounded by the top aeronautics experts, and that could give him the motivation that he required.

Musk first came into contact with aeronauts after meeting the people who were avid space-lovers who were members of Mars Society, which was an organization that was non-profit. The Mars Society was committed to studying Mars as a Red Planet and held a event to raise funds for it in 2001. The invitation was $500 per plate and was distributed to all the regular people in the community.

The only thing that nobody anticipated and that included the president of the society, Robert Zubrin, was that Elon Musk would be attending in the event he wasn't invited. Musk gave a check for $5,000, which caused everyone to pay attention. At the time of eating dinner, Zubrin made the decision to invite Musk to a coffee session to speak to him and discuss his passion for the society. He also wanted to inquire about what initiatives the society was working on.

After his engaging discussion with Musk Zubrin decided to place him at the table in the VIP section along with Carol Stoker and James Cameron, who were avid space-lovers themselves. They discussed about various space investment and initiatives. While Musk was not as knowledgeable about space, his enthusiasm for this field was very strong which is the reason Musk decided to join the Mars Society and donated $100,000!

In the course of his discussion at the restaurant, Musk decided to take one of the ideas mentioned and try something different to it. His plan consisted of sending mice Mars instead of Earth's orbit to see what they can do to endure and reproduce. He had a conversation with a friend to comment on whether this idea was insane but they didn't respond without a smile.

Musk believed that humanity had stopped innovating to the limits of real-world reality. To calm his anxieties He decided to look through NASA's site and check out the advancements they made. In the absence of any type of plans that could indicate a ongoing project He found Bupkis.

At this point, Musk had amassed quite an amount of knowledge regarding space, and had an extensive network of contacts within the business. He began his space travels in conference rooms at hotels. He would invite friends to hotels, airports, and many other locations to discuss.

It was not a business strategy in the formal sense. He simply wanted to discuss the "Mice to Mars" strategy and to see what other fans would think to the idea. The only thing he wanted to do was create a big gesture that would attract the attention of the world. The cost for his concepts was around $20 million and luminaries and scientists came together to come up with a method to achieve it.

Musk was a positive person with something positive happening, so he quit as a member of his position at the Mars Society and created his own, dubbed "Life for Mars Foundation." The number of people who attended the foundation's meeting was impressive. There were James Cameron along with some other stars, scientists at

NASA's Jet Propulsion Laboratory, Michael Griffin and many others.

Every person was a specialist in their own field of space. Griffin is the best knowledgeable of them all, with a variety of qualifications to his name. If there was a person who wanted to launch an object into space that person was Griffin. The most impressive thing was that, despite his remarkable experience, he decided to join Musk at his new company as the space guru.

The year 2005 was the time that Griffin became the director of NASA which was delighted to have someone to be able to contribute to the idea of space, as they were completely committed to the idea. In both cases, Musk as well as Griffin were enthusiastic about the idea of sending mice into space, and then observe them hum to earth. This isn't exactly high-quality television for us however for Musk the idea would be gold.

When the concept began to develop the team realized that in the process, the idea had taken an entirely different direction.

They were now putting together ideas for a project known as "Mars Oasis." Based on this idea, Musk would shoot a rocket, which would be a greenhouse , into Mars. Musk along with a team of researchers began to formulate plans to produce oxygen from Mars. There were many technical aspects involved. For this motive, Musk wanted a rocket equipped with a window as well as cameras to record and transmit video feedback. At first, some were skeptical about the concept. However, as the process progressed, everyone began to become interested. The enthusiasm of the participants was beyond comprehension and they even stated that this plan could be Musk the Time magazine's "Man of the Year."

The sole obstacle to the plan was the budget. Musk has set aside about $30 million for the project however, this alone could be wiped out with the launch of the rocket. There were many other things that were not running as planned, but it all came down the rocket. Therefore, Musk decided to take the trip to Russia to purchase a used missile. Because he didn't have any contacts within the country and he was unable to contact anyone, he determined to get in touch to Jim Cantrell, who had performed some work for the federal government from the United States. He was placed on residence arrest with Russia because of an arrangement that went wrong and Musk felt within his soul the fact that Cantrell was the person who could help him get the things he needed.

When Musk made contact with Cantrell and made an unsavory deal, Cantrell thought that people from Russia wanted to take him down. He then took him to a location where there was no way to carry a firearm. The moment they met, all their doubts were solved as Musk and Cantrell met up.

To cut a long tale short, the arrangement with the Russians ended in failure. When they returned, Musk turned to his team and stated, "I think we can construct the rocket on our own."

Cantrell as well as Griffin had too much alcohol to mock Musk therefore they simply gazed at him like a millionaire who'd confused. But when Musk revealed their plans for the rocket, they were astonished. Musk had outlined the price of components required for the construction of the rocket, as well as predictions on how effective it would be in space.

A lot of people believed that Musk was a millionaire who spent an enormous amount of money to build rockets. However, if engineers could not come up with anything with their work and he decided to shut off the work. But, Musk was dedicated, and needed someone on the team that was passionate about rockets and space. It was in John Garvey's workshop when Musk was introduced to Tom Mueller. He was a typical youngster from Idaho who dabbled with machines, and even won two science fairs in order to make his way to the top leagues.

Musk and Cantrell talked The young man eventually joined Musk's team of experts in space. This was the time Musk made the decision to start his own space company. Griffin wanted to move to another location to another location, but when Musk rejected the request of Griffin, he resigned from the team. After a few months, Cantrell left the team declaring that the entire venture was too risky and expensive.

in 2002 Musk established Space Exploration Technologies, SpaceX for short. Plans were already in place as well. Musk along with Mueller worked on rockets at the time that Musk's world was upended. Within a couple of days after his wife had given the birth of their son, Musk was shocked to discover his baby dead inside the infant's crib. The loss was devastating and, due to Musk's past of trouble it was difficult for him to grieve. Musk's wife was devastated and began IVF treatments right after. The couple then had twins, and later a set of triplets.

At this point, Musk had gotten things under control, and was able to throw himself into the everyday grind of SpaceX. The original deadline for the launch of spacecraft was scheduled in 2002. However, due to a variety of technical issues that delayed the launch until the year 2006. While the rocket flew through high in the sky, Musk watched from the control. Within a couple of minutes the rocket caught on fire and began to spin. After a few minutes, the rocket flew to the launch pad before breaking in pieces with a few of the pieces sinking to the bottom of the sea. A number of engineers were wearing diving equipment to retrieve the fragments and then put them in crates.

The blame for the launch failure was put on an unidentified crew member, who was blamed for the failure because that he did not tighten certain of the screws correctly. After a thorough investigation it was discovered it was Jeremy Hollman, a young engineer from Musk's team was last involved in the rocket. He was summoned to the office, where he and Musk engaged in a shouting contest. When the parts of the rocket were examined, it turned out they were covered by salt from Kwaj's salinity. It was one of those events that occurred and could not be detected even after a series of tests. Today, many engineers feel guilty about the way Hollman and his fellow team members were treated.

In 2007, a second rocket was launched and it performed as expected for a couple of minutes. However, this time the propellant was to blame and the rocket went into flames. The incident did not deter Musk and the entire procedure of making rockets was re-started. In the meantime, Musk had several projects in the pipelinethat we'll talk about in a bit in the near future.

It was in 2012 that SpaceX had made it to the top of their game with they launched their Falcon 9 Rocket was launched. There no issues and no parts exploded in flames. The spacecraft went on its way to the International Space Station that carried equipment weighing more than 1,000 pounds to the astronauts on board.

In 2013, a second spacecraft was launched to connect to Earth's orbital part, and then match its rotation. In the meantime, SpaceX had found its speed, and also launched other spacecraft in 2015, 2017 and in 2018.

In the last quarter of 2017, a new design for Musk's BFR that Musk affectionately referred to as "The BIG Fu*ing Rocket," was released. The BFR was built to transport more than 100 persons to Mars. Musk's timeline for the project's completion was set at 2022. In a press conference, Musk announced that the BFR will conduct a few temporary flights. In the following, he spoke about previous delays to his venture which he believed wouldn't occur this time around.

In the year 2018, SpaceX got permission from the US government to launch multiple satellites to reach the low orbit, and also provide internet service. The satellite was called Starlink and was later upgraded to the most reliable broadband service available to rural areas.

## Chapter 4: Story Of Tesla

It all began in 2003 when the experiment began. A chemistry expert known as J.B. Straubel would come up with bizarre ideas every now and then and would conduct experiments that would put his life at risk. The two-inch long scar that he had on his face is proof of the risk of his experiments. His primary interest was the use of solar energy in vehicles. The time was when no businesses worldwide had ever come up with the idea.

In 2002 Straubel was not just attempted to build his own electric vehicle, but was also moving from one motor manufacturer to another. In one of these time shifts Straubel was introduced to his Stanford group of solar cars. They were developing a plan to develop an electric car that could run on solar energy. Even though Stanford tried to shut the project, students insisted that they needed no assistance from Stanford. Straubel helped them construct vehiclesand take part in races. Straubel even went so in the direction of inviting them to his house to stay.

An evening conversation following the race, they were talking about how the lithium batteries been more effective in a car. He presented his idea of lithium battery to Stanford group and they all agreed that they'd be willing to take it on should he come up with the money for the project.

Then, Straubel stalked investors for cash in shame. He was campaigning that Harold Rosen took Straubel with him to speak with Musk. Straubel pitched his concept of an electric plane, but received none of the response from Musk. But, when he spoke of electric cars, his idea caught the attention of Musk.

Musk himself was considering the possibility of creating electric vehicles for a long time. When he learned of Straubel's plan to use lithium battery the idea was his to be believed. Musk offered Straubel 10% of the $200,000 he had been looking for and forged a wonderful connection with them due to their common interest.

To impress him even more, Straubel asked the President of AC Propulsion a tzero for

Musk to drive. Musk was enthralled by the car and even tried to make it commercially available. But the engineers at AC Propulsion wouldn't agree to the arrangement. Musk later decided to pursue his own route.

In the background, Straubel did not know that Two business associates Martin Eberhard and Marc Tarpenning were working on their own idea of electric vehicles. The year 2003 saw the forming of a business was founded, called Tesla. The owners, as well as others who invested, had the abilities and know-how to manage a business and also to make automobiles. The next step was to get support from someone who could know how to handle such projects. In their heads there was only one name Elon Musk.

Because Musk was already searching for an entry point his company, he was eager to meet Eberhard and Tarpenning. After hearing the things they were saying the two of them were on board. Following this meeting Musk promptly called Straubel and asked him to meet the men. After that, although Straubel was doubtful because he'd not heard of Eberhard and Tarpenning however, he nevertheless paid them an appearance. He shared with them the batteries he was working on and Straubel was appointed on the spot.

Musk was now collaborating with business partners, and he had hired an ambitious team of engineers. The team began working to build "Roadster," the first vehicle that Tesla was planning to build. They were unaware of what was to come for them. In the meantime, Musk faced many issues. The majority of his engineers were working in Kwaj and engaged in developing the Falcon in SpaceX. Tesla has now grown into a massive company, that had hundreds of workers. The creation of an automobile that could compete with the best was not an simple task. The engineers had to deal with similar problems to those they faced in the beginning with Falcon.

The employees at Tesla witnessed a different side to Musk when the original Roadster was a failure. Some were verbally abused, thrown out of meetings, and then fired in a flash, including employees who had performed excellent work before. There was no one safe from Musk's scorn. In a personal sense, Musk was having a hard time to deal with Justine as well as her website as well as on a professional scale it was difficult for him to find a way to stop his creations.

In this period, Musk also became an source of inspiration for the film Iron Man. In the event that Robert Downey Jr. got the news of a man that was very similar to the role model Tony Stark, he decided to visit SpaceX. SpaceX headquarters. They had a great time and Musk enjoyed having a chat with somebody who knew a bit about technology. In the beginning, when Downey began filming in the movie, he asked Jon Favreau, the director of Iron Man, to place the Tesla Roadster in the movie. On the surface, he was trying to show that Tony was an amazing character and had a Roadster before the movie was released. Musk started to appreciate his relationships with celebrities and Justine and Justine received invitations to numerous Hollywood occasions and even weddings.

The search for the perfect Roadster was never over. In 2008, Tesla finally showed off the Roadster that was capable of reaching 60 mph in 3.7 seconds. It could also travel up to 250 miles.

At this point, Daimler had a stake in Tesla and Musk had signed a deal with Toyota. The company's initial public offering began in 2010 and brought in $226 million. It was in 2008 that Tesla unveiled a brand new car, called Model S. Model S, which would be the first electric car developed for the firm. The car finally came out during 2012 and 2013 as well. Motor Trend Magazine awarded it with the title of "The car of the Year" for the year 2012."

As of 2017, Tesla has surpassed General Motors and was the most valuable car manufacturer across the US. Musk was extremely impressed by the progress made and accelerated production for Model 3. Model 3 sedan. At this point, Musk was on a good run and things were progressing very well. In the year 2019 the Model S four-door sedan set the fastest speed record at the time in Monterey County, California at the Laguna Seca Raceway. It was revealed that the Model 3 sedan was finally released to the general public in 2019 at a cost of $35,000, significantly less than the $69k price tag that was offered for Model X and S electric sedans.

In the year 2019 the Semi-Truck was revealed after numerous delays. Later that same year Model Y, Tesla's long-awaited credit card, was revealed. It was on sale in 2020. The model was able to reach 60 mph in 3.5 seconds and had 300 miles of range. The range of these cars isn't over. Tesla models, and Musk promises to introduce new cars with advanced technology that will help people live their lives more easily.

## Chapter 5: The Musk Who Is More Than Just Tesla As Well Spacex

Musk is more than simply an e-mail address. Musk is a brand that promises the latest technology. PayPal, Tesla, and SpaceX aren't the only companies Musk created, sold and manages. Musk also has smaller companies that people aren't aware of like SolarCity that was purchased in the year 2016. The company's mission is the advancement of renewable energy sources.

He also founded The Boring Company in 2017. The company is building tunnels that reduce street and road traffic. There are many inventions that bear his name too Some of them are in development. The public can learn more about them on the internet. A majority of these inventions are aimed towards improving the quality of life for humans.

In these tales One of the things you've discovered was that Musk is an avid runner. Musk believes in doing things according to a plan, and occasionally it is just a matter of jumping into. His ruthless and controlling personality has brought about many problems including being removed from the position of CEO. But with his hard work and commitment He built a new company from scratch that was far superior to earlier ones.

Presently, Musk is constantly in the news, not only due to his wealth, but also due to the empires he's created. Furthermore his plans to conquer Mars and his ongoing talk about how AI could dominate the world and use animals as pets does not go over well with many people.

## Chapter 6: The Hidden Secrets To Elon Musk's Achievement

"Failure is a possibility here. If you're not seeing results or you're not making enough progress."

If you think of Elon Musk, one of the top ingenuous billionaires on the planet the name Elon Musk comes to your thoughts.

Disruptor. Trendsetter and game changer... These are just a few of the words that are associated with Musk. The most often talked about achievements include Tesla, SpaceX, and PayPal. For Musk every workday is spent trying to launch a rocket or creating a Tesla model.

Musk is the real Iron Man, who commits his entire self to the concepts that are in his head. In contrast to other billionaires, Musk has been able to lose his money and earned billions more and this was repeatedly in different companies.

In 2021, Musk was the wealthiest man on earth and he is in competition alongside Jeff Bezos and Bernard Arnault. Musk has been named on in the Time magazine's 100 Most influential people. Musk and his popularity increase each time he thinks of something innovative. So, what's the secret?

Find out more:

Give Your Best and Do Your Job as hard as you can

"Starting and growing a company is more about the creativity as well as the drive and determination of the people who run it as the products they offer."

Musk's drive to succeed is ferocious. When he and his twin brother were employed at PayPal and lived at the workplace. They ate there, rested there, and had baths in the YMCA. They were so committed to building a successful business that there was no limit to what they could do. They were incredibly hard-working and worked 100 hours a week.

Musk believes the dedication he showed to his work increased his chances of success. His motivation behind his determination to succeed was that If people work 40 hours, and you're working twice as long, then you can achieve your goal in just 4 months, while it takes others 1 year to accomplish the same goal.

Make sure you are passionate about what you Do

"People perform better when they understand what the objective is and what it is for. It is crucial that people enjoy coming to work every morning and are able to enjoy their work. Do what you're interested in. It will make you more happy than just about everything else"

Each idea Musk had in mind was a result of his fascination with the subject. As a bank employee and financial institutions, he understood the difficulties they faced with debt. This inspired him to establish PayPal. With SpaceX being a part of it, his fascination with Mars was the main reason which led into producing green energy. The ventures that were launched at random and carefully studied and were supported by a passion.

Do not follow the trend

"I don't start companies to create businesses, but rather to get things accomplished."

For Musk the company was never about the creation of a brand. Musk's passion was developing solutions for today's world issues. He is a pioneer in the field, and his philosophy is not following the latest trend. Musk's businesses tackle problems, and compete in fields with less players or none even.

When PayPal was first established it was the first company to allow users to transfer money via email. SpaceX is the first company to launch its spacecraft Dragon on its way to ISS. Finally, Tesla is the best and largest electric car maker worldwide.

To reap greater and more rewarding benefits, you must be focused on innovation. There will be less competition when your concept is distinctive. When there's a monopoly there's plenty of room to be creative, especially if profits are greater.

Do Something Important

"I always believe in optimistic views, however I'm also realistic. It wasn't with the hopes of massive success that I founded Tesla and SpaceX... It was just because I believed they were worthy enough to pursue them anyway."

Musk was not the one to start SpaceX or Tesla simply because he came up with an ingenuous idea. Musk was deeply worried about the direction that our world was headed. He was aware that if people kept taking the resources that they had as a given, nobody could ever advance as the coming years would be identical.

In his 20s, Musk earned more cash than he ever thought of. Musk could have easily retired and had a comfortable life. But he and continued to work. For him, work was of more importance than making money. It was about changing the world. So, take a risk and pick something that is significant to you.

Read to Lead

"How did you get started how to build rockets?" Musk: "I have read books."

Self-education never ends! It's impossible to know every aspect of life, but you have the ability to attempt. When he was a child, Musk read the whole Encyclopedia Britannica. Today's generation is making memes and reading social media posts, Musk's young self-set a reading pattern.

The wise realize that getting lost in a book is the most effective method of expanding your perspectives. Reading will not only give you fresh ideas, but it will also provide knowledge that will alter your way of thinking. Musk is of the opinion that a quick learner is an excellent earner.

Don't be afraid to speak up about constructive Criticism

"I consider it vital to keep a feedback loop in place, that you're constantly contemplating what you've accomplished and how you can do it better."

Although Musk was well-known for his temper that could have any employee afraid for their job, he certainly was not the type of person who was quick to act with no thought. Sure, he made a lot of mistakes while SpaceX was being built including firing engineers that weren't in the wrong. But, in the pursuit of success, he'd do anything, which included accepting constructive criticism. If a situation was discovered on his watch, he sought advice and guidance about how to solve the issue.

Many CEOs and others in leadership positions don't like to hear about what they're doing wrong. They consider it an insult. However, Musk always sought valuable guidance from trusted individuals and his friends. Prior to launching a business idea that he was sure people would react to with suspicion He would inquire from his peers whether he was crazy to think of something like this.

If the negative feedback or criticism is based on an explanation that doesn't make sense inform people to alter their behaviour. Musk admits that he's not an expert, but requesting negative feedback, something that a normal person would never do, is the way Musk operates. According Musk, if the suggestion is from someone close to you, they're in your best interests in mind.

Attract the Best People

"My most erroneous mistake is probably placing too much emphasis on someone's aptitude and not their personality. I believe it is crucial to determine if one has a heart of gold."

Musk continues to think about the moment that Zip2 was under his control. Zip2 was removed from Musk. Also, there was the time that an attempt was made to get Musk from PayPal. The most important thing Musk discovered during these endeavors was to don't trust anyone based on their talents. Sure, they'll do the job more efficiently than others, however, do they possess a strong heart? Are you aware of their motives? Do you think they'll hit you in the back whenever they see?

A company is formed by a team of individuals committed to a single purpose. If they have the same goal, figure out if they're also loyal. Musk was a victim of being in a advisory position in his first two ventures. In the following years, he owned the largest stake in the company, so that nobody could force him out.

Make a decision to take risks, or else life is going to get in the way

"I could just watch it happen, or participate in it."

Have you family members told you that there is an opportunity to build your own career, but that family first?

Musk believes in the reverse. He believes that once you have an extended family, it's difficult to make a decision and the people you love find themselves right in between you and your desires. There is no way to escape obligations. If you wish to succeed in your profession it is important to decide for yourself first!

Don't think of this as an act of selfishness. Certain things can cause you to slow down in your life. Therefore, you should take a chance and choose what is the best choice for you. If it means focusing all your time and energy on your work, then do it.

Be Extremely Tenacious

"I do something and then it typically occurs. Perhaps not exactly on time but it will happen."

One of the biggest mistakes made by Musk was the fact that he did not comprehend what people were trying to communicate. Musk's focus led him to not understand the other events that took place right under his nose. Musk

Believes that hiring a pleasant person is essential to running a business because they are just as determined as you are.

If you go to"tenacious" in the dictionary, you will find "tenacious" within the Dictionary, you'll be able to see it defined as an adjective: not giving up a position, principle or the course of action that you have decided to follow.

Even in the midst of the fact that Tesla as well as SpaceX were on the verge of becoming bankruptcy, Musk stuck to his goals and made as much as could! He took the money through PayPal and put the bulk of it into both companies to ensure that none of them experienced any problems, which they already had.

Sure, Musk could be a bit stern and sometimes rude. But, his ideas weren't just out of on the whim of his mind. They were thought-through using graphs and statistical data.

Take a look at what the worst thing that Could occur

"Persistence is crucial. Do not quit unless you are forced to quit."

In the early days when Musk was beginning his own business when he was just 17 years old He decided to carry out an experiment to determine how he could manage to live with as little as he could. Every day, he consumed hot dogs and oranges, which cost only $1. He earned just a few dollars a day, which meant the person he was purchasing from was making $30 by himself. That didn't seem too hard.

Musk did not do this because the man was in poverty. At some point within his own life wanted to know if he had lost all of his money and he was unable to pay it back, what would occur.

According Musk Musk that's how an entrepreneur should be that is committed to achieving it. Of course, he's achieved success and earned millions of dollars. He also understood what it was like to be poor and this thought was reflected in the business decisions he made. He was never a part of any venture he wasn't sure of in. Furthermore, he didn't approve the purchase of something unless it was absolutely necessary.

Sex Sells

"What is it that makes you think in a new way? ?... I believe it's the mindset. You must decide."

Say, what?

You may be thinking about what sex has to have to do with any of Musk's inventions. It's true that the sex industry is lucrative and if have the right strategies for marketing

You can utilize this trick to get your business to the top.

If you glance at Tesla automobiles, you'll find that they all contain sexual undertones. Musk is not just a genius but also has a unique style of comedy. Tesla launched three versions of its cars one after another with distinct models with different names. The first was S followed by X and the final one is Model 3, which was depicted by 3 horizontal lines. The model was initially known as E. Combine the names of the models together and you'll have one word SEX.

In the following days, Tesla introduced the D dual motor. At the time of its unveiling, Musk made fun of the new model while in the background, one heard a fan say, "Show us your D" to which Musk smiled and replied, "You will notice my pants are sewed with Velcro stitching."

Put Your Money Into Profits

"The initial step should be to prove that something is feasible; after that it is likely that it will happen."

Musk has not hesitated to invest his money into ventures Musk believes in completely. The two businesses he founded were sold with millions in cash and every time, he took at most 45% of his earnings to invest them in an entirely new venture in the following year.

* In the sale of Zip2, Musk made $22 million. Musk used $10 million to discovered X.com that was later renamed PayPal.

* In the case of the sale of PayPal, Musk made $165 million. Musk used the money to discovered SpaceX.

* Musk also invested in SolarCity (20,724,991) and Tesla Motors (33,076,212).

The owners and control SolarCity, SpaceX, and Tesla are the three businesses that contribute to Musk's billion-dollar fortune.

Develop your ideas to the point that You're Considered to be the Best

"There are really two factors which must occur to allow a technological advancement that is affordable. One is that you must have economics of scale. Additionally, you will need to test the concept. You will need to try several versions."

The year 2015 was the most popular, Consumer Reports tweeted, "Breaking The Tesla P85D is the most amazing car we've ever tested over the past the past 80 decades of testing vehicles." #carofthefuture

This is the kind of acknowledgement that you will need to increase the image of your company. To get to this level, you have to give your best everything to your product and Musk did just this.

Be aware of your limits

"Don't think that things are working, when they're not, or you'll be enthralled by a wrong solution."

Musk is the most difficult and smartest entrepreneur working in today's 21st century. But, he has a life and can only be productive for a certain amount of time. With five kids and his wife, he must to split his time between wife and family as well as work. Tesla and SpaceX consume the majority of his time. Musk was aware that launching another business would be too to handle for him.

Although he was filled with ideas, some of them quite excellent ideas that could create lots of changes in this world, the founder chose to hold off before jumping all his feet. When he started SolarCity He was aware that he would not be in a position to supervise its operations each day. Therefore, he discussed this idea with brothers Lyndon as well as Peter Rive, who agreed to oversee the business. Because Musk was repeatedly sacked before, Musk realized that he needed to have some sort of control over the company. And that's why he became the primary investor.

In the age of digital technology that is the age of automation where human workers are replaced by robotics, artificial intelligence machines and algorithms as well as automated processes It is obvious to conclude that Elon Musk is the utmost person in today's Twenty First Century even though his significance is largely surpassed by Jesus Christ, the Jesus Christ, the Son of God and Messiah Jesus Christ. Elon Musk is a billionaire with an estimated net worth of $23,000,000,000, is the director of SpaceX and Tesla Motors, and knows how to think beyond the bounds of the metaphorical boxes, as a pioneer from the future. Elon Musk's ingenious technologies have been developed by him are arguably the most important figure in the age of digital.

Elon Musk has fundamentally transformed the world through his revolutionary inventions and groundbreaking ideas. These breakthrough technologies will help humanity and allow the world's civilization to live in an environmentally sustainable wealthy, abundant, and sustainable future. " Between Musk's work at Tesla making electric cars and his ambitions to place humans on Mars in 2024 (and in the future, be able to colonize the planet) The millionaire executive in the tech industry is working to transform human transportation in space and exploration. Elon Musk is trying to create a new future and not only by creating the next application however, but by taking on global problems. Elon Musk has the potential to have the biggest long-term impact on our society. Elon Musk will change the way that civilization operates as we know it " (Huddleston Jr. in 2018,). Elon Musk has ushered in revolutionizing technological advances for the entire world.

His inventions by Elon Musk have been innovative and are changing the way we live in the digital age. For instance, Musk's Tesla Motors company has produced electric vehicles that can be autonomous. "There is a lot of need for electric vehicles particularly models like the 2019 Tesla Model 3 (which boasts 200 miles of range and is priced at $35,000) however, Tesla Motors is still struggling in producing enough vehicles to meet that demand" (Morris 2015). SpaceX, the Elon Musk-owned Aerospace SpaceX company SpaceX is developing rockets that can be reused. SpaceX was established "to aid in reducing costs of space travel by 90% and allow to colonize Mars. SpaceX's Falcon 1 and Falcon 9 rockets are engineered to be reusable , and SpaceX has launched six cargo resupply missions for SpaceX's

International Space Station" (Morris 2015). SpaceX is the company of Elon Musk. SpaceX can make space exploration more efficient and economical by introducing innovations to the reusable launch system interplanetary transport systems, rocket engines spacecrafts, satellites as well as launch vehicle. Elon Musk understands the salient importance of making humanity an interplanetary being that is not solely dependent on the natural resources Earth can provide for our survival.

The rise of space mining as well as mining asteroids is the catalyst for the destruction of the silver and gold bullion market for the typical investors in silver and gold. Although the prevalence of space mining may appear absurd in the current digital age and the technology of automation is feasible with the advent automated mining robotics in the workforce as well as the sweeping advances in aerospace engineering technology that can make space exploration more economical viable, efficient, and efficient. The innovations that SpaceX is its Aerospace firm, can made to bring into the sector of space exploration will possibly make Elon Musk the most significant figure to be a part of this Twenty First Century who will be remembered for centuries to come for being a pioneer into the future of automated technology because the space exploration he pioneered has been feasible efficient, cost-effective, and efficient.

Elon Musk profusely understands that space could also be the ultimate frontier for countries that depend on oil. "Saturn's bright moon Titan contains thousands of times more hydrocarbons in liquid form than the natural gas and oil reserves on Earth as per new Cassini data. Hydrocarbons pour down out of the sky, and are deposited huge deposits that create dunes and lakes" ("Titan's organics on the surface," 2008). Moons and planets from other stars could be unexplored gold mines which are rich in reserves of oil and various other resources from nature that we are unsustainable decreasing of the Earth.

The space explosion and the boom in asteroid mining could not just enable companies that are not part of the space mining or mining asteroid industry to create enormous wealth in the era automated technology, it it will likely also bring about an era of prosperity in which depletion of natural resources is no longer an apocalyptic problem. Space mining's ubiquitousness will provide the chance for humanity to bring huge quantities of oil as well as various other resources from the Earth that we are currently draining of the Earth. With the abundance of extraterrestrial planets, moons, asteroids and moons to extract resources from the economic consequences could be life-changing and bring about a new period of prosperity, especially when these mining operations are performed by sophisticated robots that can import resources into Earth throughout the day.

The impact of mining of space and asteroid explosion on economics will prove significant, as the space mining and asteroid mining boom will allow countries to produce immense wealth and usher an age of abundance which the scarcity of resources from nature is not a serious issue.

"It is believed that some asteroids may contain all of the platinum mined from landmines across the past and could have a value in the hundreds of billions of dollars. In reality, it's assumed that some are containing cobalt, nickel, or iron in sufficient quantities to satisfy the requirements of Earth for the next 3,000 years. There are around 12,000 asteroids passing each year near to earth in a variety of sizes, from huge rocks to pieces with a diameter of several kilometers. It would also be much easier to get on the 10 percent of them rather than the Moon" (Vyas 2019). The consequences of mining of space and asteroid boom could change the course of our lives and beneficial to the industries that benefit from this boom.

The innovations in technology of Elon Musk's SpaceX company brings to the aerospace engineering field will open openings for mining of asteroid and space minerals explosion. SpaceX's revolutionary technology will ensure that space mining and asteroid mining feasible and cost-effective in this time of automation.

Elon Musk is not just co-founder and CEO of X.com and the majority shareholder of PayPal He is also the inventor of the innovative Hyperloop transportation system. "The Hyperloop transportation system aims to enable commuters to journey from Los Angeles and San Francisco within 35 minutes or less, which is much faster than a regular flight. This innovative rail system would operate in a tube that has very low air pressure that could reduce drag, and permit faster speeds of up 800 miles/hour and less energy consumption" (Morris 2015).

Elon Musk also co-founded Zip2. "Zip2 was designed to help the publishing industry create "city guides" for users of the internet. Zip2 was purchased from Compaq to the tune of $307 million, in 1999. Musk was paid $22 million through the deal" (Morris 2015).

Elon Musk also co-founded the solar power company SolarCity. "SolarCity has expanded to become the second-largest supplier of solar-powered systems in the nation with revenue at $102.8 million during the last quarter. It is currently in the process of developing storage systems that allow people to draw power generated by solar panels during the night. The company is cooperating with Tesla to provide free solar-powered charging stations for owners of vehicles travelling Route 101 from San Francisco to Los Angeles or back" (Morris 2015). The Elon Musk's SolarCity company is constantly making progress toward bringing his concept of the future powered by solar to the point of.

Elon Musk achievements and ideas go beyond the previously mentioned. "Elon Musk has also invented the idea of using web-based telephone calls that let computers call landlines. The concept was that people could simply click on a company's contact information online and calls will be directed to the company through the phone center" (Morris 2015).

Furthermore, Elon Musk thought of search results that are specific to locations, which could make commerce more efficient. "His idea was to design the system to initially search for results in the geographical zone closest to you before automatically expanding the area (meaning users would not need to refine their results) to get enough outcomes" (Morris 2015).

Elon Musk also ambitiously plans to provide high-speed internet access to millions of people at reasonable prices by using satellites. "SpaceX's program, dubbed Starlink is based on the idea that a "constellation" of satellites can orbit Earth and transmit Internet back to the ground at high-speed speeds. If the project is successful, it will bring Internet to the millions of people who don't currently possess Internet access. It could also offer an affordable alternative for developed countries that have Internet access via cable or wireless carriers. The company is authorized to launch 12,000 satellites across space over the course of time. When that time comes the network will be set up and Internet access will be made available throughout the globe. The full-fledged satellite Internet service operating in the low-earth orbit" (Reisinger 2019) could revolutionize the world as the internet offers employment opportunities, education opportunities, and wealth-generating opportunities, and the means to communicate globally.

There is no doubt the fact that Elon Musk has been deemed the single most influential person in the modern age principally due to his groundbreaking inventions that will propel humanity into a new age of prosperity and enable people to attain the highest quality of life. Older inventions, such as automobiles satellites, rockets, and ships are being redesigned updated, upgraded, and modernized thanks to the innovations of Elon Musk. Elon Musk is focused on creating interstellar space travel, interstellar colonization as well as space mining, the mining of asteroid, and satellite-based Internet popular in the coming years. Elon Musk is a pragmatic solution provider that aims to create a sustainable environment through homes that are powered by solar power and the creation of resources brought to Earth from spacecrafts and planets. Elon Musk envisions a world where vehicles are eco sustainable, extremely efficient electric, autonomous vehicles. It's not difficult to comprehend why even futurists and theoretical physicists consider Elon Musk as the most significant person in the digital age.

Elon Musk is set to transform the world in a profound way. The companies of Elon Musk will bring into a revolution in the world, which will improve the globalization of our civilization, improve our technology, and dramatically raise the standard of life. Elon Musk's companies have already transformed the world and are now in their infancy stages of technological technology. Elon Musk's radical transformations to our world "have already taken place. From PayPal becoming the dominant world of digital currencies to SpaceX already launching flights, to Tesla developing an affordable electric automobile (with auto-driving capability) Musk's innovations and companies have revolutionized several aspects of how we do business and live our lives" (Bryant 2019, 2019).

Elon Musk revolutionized the commerce by streamlining the checkout process through the introduction of PayPal. Elon Musk was "a co-founder of PayPal Holding Inc. PayPal has already revolutionized how we pay. You can sign up PayPal and make a transfer to the brother to pay for dinner the night before. You can buy something from a company located in China. You can enter your PayPal information when shopping at Home Depot and not even think over bringing in a credit or debit card to the shop. In a nutshell: PayPal has been able to take over the digital currency world and is so embedded in our brains that we can't recall a time without it" (Bryant 2019,). PayPal provides a secure and easy option to pay online and transfer funds.

Elon Musk revolutionized space travel by launching SpaceX, his aerospace firm. SpaceX. As we've mentioned before, Elon Musk has rendered space exploration efficient and cost-effective by introducing innovations to reused launches, interplanetary transportation systems rocket engines, satellites, spacecrafts as well as launchers. Elon Musk plans "to make sure that Mars can be inhabited in the event that the earth becomes an option, and is planning to send humans into space to establish colonies on different planets" (Bryant 2019). Elon Musk will presumably play the leading role in drawing an additional gold rush, in where asteroids and plants are mined for minerals through mining robots. The financial viability of mining space and asteroid mining missions will trigger another gold rush take place out in space, not on Earth.

This second gold rush which will occur as a consequence of asteroid and space mining missions that are performed by mining robots will make the resources of precious metals like silver and gold, plentiful and not as readily available. This will result in an unprecedented decline in the value of silver and gold bullion as precious metals can not be considered to be as scarce resources. This will ultimately end the gold and silver bullion markets for the average bullion buyer, because prices for silver and gold bullion are likely to fall to record lows in real dollars.

Space mining and asteroid mining which are carried out by robots mining The changes in the prices of silver and gold bullion may result in them becoming more volatile than changes in the price of precious metals due to prices being manipulated and suppressed, which has caused the prices of precious metals, like the gold price, to become massively diminished. It is plausible to consider that the rise of mining in space and on asteroids could destroy the bullion and gold market for the typical investor in silver and gold who believes the precious metals at a minimum, have the same worth as they do now in the future time of abundant resources in which precious metals will accessible in abundance. Elon Musk's inventions in space travel with his SpaceX company SpaceX might not only make space mining and asteroid mining more cost-effective and viable however, they could also play an important role in the next few decades in allowing the scarce natural resources to be readily accessible.

Elon Musk's business, Tesla Motors, has introduced autonomous electric vehicles on markets that's more secure and greener for driving than gas-powered human-operated vehicles. "Musk's intention when he founded Tesla Motors was never to develop a vehicle powered by electricity that only the wealthy could be able to afford (the Model S starts at $70k, while it's Model X, an SUV priced at $80k) However, the goal was to provide electric vehicles that everyone can be able to afford. However, since driving a car that is powered by the grid (electricity produced from burning coal) will only transfer the pollution elsewhere, there is a higher goal" (Bryant 2019,).

The company of Elon Musk, SolarCity have not only changed the way that homes get electricity, but also offers an affordable, eco green option for powering the home. "SolarCity is determined to create renewable energy that every house is not able to afford, but also want to include. SolarCity creates and installs green energy systems for both commercial and residential environments. They've designed their own Tesla Battery, a high-efficiency stylish battery that can be enough to get someone off the grid (given sufficient solar panels, a low-consumption and other elements). The idea behind this is that with SolarCity's green energy upgrades, householders can become much less dependent upon electricity and charge their electric vehicle and eventually reduce their impact on the earth without spending a lot of money" (Bryant 2019, Bryant). Tesla Battery Tesla Battery has made the cost of utilities for homeowners much less expensive because solar energy is much more affordable than the use of electricity generated by power plants.

Elon Musk's Hyperloop transportation system was designed to revolutionize transportation and solve the issues with freeway traffic. "The Hyperloop's transportation system looks like something straight from Futurama. Passengers would step into a two-meter wide pod, which would then be zip-lined through a long tube. The whole thing would be powered using (presumably) renewable electricity. It would result in an environmentally friendly method of transport that can ease the burden on traffic across the nation (when placed in other locations). Musk says that his Hyperloop could cost only $6 billion and will be capable of making the journey within thirty minutes between LA up to San Francisco" (Bryant, 2019). If the Hyperloop were to be implemented across the globe, then congestion-ridden highways and lengthy travel times would not be a problem in the age of digital technology.

Elon Musk has profoundly altered the world and his business ventures will continue to transform the world in the coming years. "Elon Musk is bringing science fiction fantasy closer to reality than most people think. His ideas and innovations can transform and improve our lives. the lives we lead" (Bryant 2019,). Elon Musk will continue to not just transform the world over the next years with his company's technological advancements, but also enable humanity to live the most technologically modern, prosperous, and resilient future.

The Elon Musk-owned SpaceX aerospace firm has created numerous innovative technologies that could transform and change the way we think about space travel. The most important of all, "SpaceX's Falcon 1 became the first privately owned liquid fuel rocket that launched into earth orbit. It was the first to launch commercial satellites through the launch phase to its orbit" (Williams 2018, 2018).

In the second place, SpaceX was able "to make the first ever commercial spacecraft to fly as well as to and from space, including the International Space Station with the use of their recycled Falcon 9 rocket. SpaceX was the first private enterprise to make the first orbital class rocket using their recycled Falcon 9 that was successfully returned to space, and then landed in one of SpaceX's drones within the Atlantic Ocean. It was the Falcon 9 rocket also achieved the first landing on a straight plane of an orbital first stage rocket on land" (Williams in 2018,).

Third SpaceX's Dragon rocket was able to achieve the feat of becoming the "first private spacecraft" to travel to the International Space Station. SpaceX's Dragon rocket cargo spacecraft transported to and from the space station with no problems. Following this accomplishment, SpaceX has sent the reused Dragon rocket to orbit several times in order to complete resupply missions" (Williams 2018, 2018).

Fourth fourth, SpaceX's Falcon Heavy rocket achieved its first "heavy test flight. Following a success in its launch Falcon Heavy is Falcon Heavy has become the strongest rocket currently in operation in the entire world. The payload model is Musk's Tesla Roadster sports car, featuring a dummy astronaut named Starman as its driver" (Williams 2018,). SpaceX's Falcon Heavy rocket has an incredible payload capacity of 140,700 pounds. The Elon Musk-owned SpaceX company has come up with new aerospace engineering techniques that make space exploration much more feasible efficient, economical and efficient.

Elon Musk fully comprehends the significance of entrepreneurship as well as solving problems that have been a century in the making to make huge wealth. Elon Musk clearly understands the need to create unimaginable economic value for entrepreneurs to make money, flourish in the long run, flourish, and prosper because the majority of corporations do not even think about paying an adequate salary to their workers.

Companies will never compromise on paying their employees an adequate wage, nor will they ever even think of paying their employees a reasonable wage for any of a multitude of reasons. In the age of digital companies have a tendency to pay their employees who perform dead-end and dispiriting, grueling job that is not fulfilling a minimum wage, which does not even offer any close to the sustenance wages they receive for housing. The majority of corporations don't offer their employees any kind of benefits despite all the toil, struggles and brutally painful labor that employees must endure in order to continue working their unsatisfactory disappointing, unfulfilling low-wage jobs. In the age of digital employers realize the fact that their employees make up a dime dozen and see their employees as unprofitable capital-based livestock that can constantly be a part of for countless hours per year for the purpose of making money.

With no access to universal basic guaranteed income or having had the necessary time required to develop income-generating assets after having been forced to forfeit thirteen years of sacred, precious period of childhood because they have been ensnared into the 13 year impure, compulsory k-12 indoctrination camp and completing thousands of unimportant homework, students are dependent on their employers as they grow older and are likely to accept an unproductive, depressing work-intensive, unfulfilling and painful minimal wage work that doesn't offer a decent wages just so they can afford to purchase some food. Also, employees do not have any real bargaining power in an employer market where possibilities for employees are scarce. In addition, the limited jobs for employees are shrinking as companies continue to work towards replacing the human workforce with robots, machines as well as artificial intelligence and automated processes.

Furthermore, companies view human workers as a risk because they cause them to be liable for costs of labor which reduce their profits even though they are the mainstay of the companies that make a profit from their work. They view their employees as a petty, capital animals that have to constantly work for thousands of hours a year to earn profits as cogs on an economic machine. It is because of this view that the majority of these businesses don't pay their employees anything over the minimum salary.

The income gap has become so massive so that, in United States for instance that 8 people own more money than the lowest 60 percent in the country. Corporates have also reached market caps that exceed $1 trillion.

In the digital age technological companies, like Apple and Microsoft have grown to become the most profitable firms with market cap values surpassing one trillion dollars. The reason why they are profitable is due to the fact that they can sell at any cost for their services as well as products that are distributed digitally, giving them the chance to increase the profits they earn. The largest tech companies each generate millions of dollars annually in net income and usually achieve profit margins that exceed 30 percent. Apple for instance, had an earnings margin of 38.34 percent as of June 30, of 2018, while their rival Microsoft posted a profitability margin of 39.11 percent at the end of June of 2019.

Corporates are able to be able to afford paying their shrinking human workforce an affordable wage, but are not inclined to pay them because their primary goal of maximizing profits, and humans aren't just their biggest expense however, they are they are also the most expensive thing in the labor market, which prevents employers from ever paying anything more than the minimum wages to workers who they view as disposable and easily replaced part of the economic system in the employment market.

They also fully comprehend that human employees are burdened with insurmountable debt and huge expenses, which implies that they are willing to take any insanity-filled, exhausting work-intensive, unfulfilling and painful minimum wage job in order to purchase food in a society where living costs have been deemed prohibitive. Employers hold all the bargaining power and will gain more influence over their employees when more jobs are eliminated and the applicant pool expands. Corporate executives are aware that their employees are in desperate need of an incredibly small amount of money and will accept any precarious, unending and gruelling painful, insatiable, minimum wage job that they can achieve out of complete desperation to ensure they will be able to survive in this fake real world.

In the end, because corporations hold all the bargaining power leverage, wealth, and power in a job market for employers that is dominated by human workers only a handful and are viewed as a labor cost burden to them, they sincerely know that they are able to find workers to do their precarious, unending and dispiriting, hard painful, unfulfilling minimum wage job , which does not pay more than the 1/4 of a living salary to afford housing. Corporate employers will never have to pay anything even close to an affordable wage to their employees because they are aware that employees are desperate to secure any insecure work-intensive, unfulfilling miserable minimum wage job because of desperation and sheer rage due to the burden of debt, massive bills and no financial resources. The people who are in debt have no chance of bargaining power in settling for the minuscule amount of fiat currency for their hourly wage rate . their miserable, dead-end minimum wage position since employees must always earn money in order to eat and purchase food that is expensive.

The main purpose of a business is to maximize its profits as well as the prosperity of shareholders. The majority of corporations view their human employees as costs that impede their capacity to achieve the highest profit possible, which is just one of the more reason why corporations offer their employees the lowest wage they can get to pay in this day of automation. Much to the dismay of the employees companies will never think about paying a decent salary to their employees because they view their employees as cost of labor assets and as exploitable, disposable capital livestock that is that can be leveraged to make money. Companies want to extract maximum economic benefit from their human workforce with the least amount of expense.

Human workers as we know is becoming a dying breed in the current workforce which is plagued with the unemployment rates of between 80% and 90 percent as a result of the lack of jobs in the private sector that offer economic value, as evidenced by the public demand for their services and goods. Boondoggles jobs that do not bring any economic value to the market shouldn't be subsidized through the salaries of private sector workers who have real jobs that offer economic value. If you look at the percentage of people working in private sector jobs that bring economic value, it is possible to determine that the unemployment rate is around 10%-20 percent.

This market system is filled with boondoggles jobs that would not be there in a free market system where everything was privatized and there were no rules that were imposed by bureaucratic systems. In a controlled market system individuals' hard-earned funds are used to finance the trust funds, the amenities like vacations, pensions and vacations. lavish benefits, extravagant salary of public-spirited welfare aristocrats and at the expense of the minimum income time, prosperity, happiness, dignity and health. There are boondoggles jobs in the private sector in a controlled market that are not real jobs because they are not because of a demand that is voluntary.

In this market that is controlled it is mandatory to carry auto insurance to use your personal vehicle, regardless of whether or not you believe that it is worth the cost. Furthermore, since you're not in control of the way your money is allocated in a market that is controlled there is also no control over your time in a controlled market.

In this case, for instance, against your own will you're required to waste 13 years of your precious limited, precious time that is wasted by being taken into a 13 year insalubrious, k-12 obligatory indoctrination camp. In this 13-year, insalubrious, compulsory indoctrination camp , you're required to finish thousands of hours of useless homework assignments which provide the student with no reward for your effort, time and efforts other than the deprivation, stunted growth, constant stress, poverty, decreased intelligence, agony stress, depression and telomere lengthening. Being forced into this 13 year impure, k-12 mandatory indoctrination program also contributes to the high rate of unemployment because people are forced to work for at minimum 13 years in their lives despite their desire to get productive.

If someone is stripped forever from their life, time liberty, wealth, and happiness , it is the conditions for unstoppable wage slavery, particularly in the absence of a guaranteed income that can provide the wage-slave with the means to escape this precarious position by having the money the expenses of self-employed ventures. Entrepreneurship that is self-employed can be extremely time-consuming to pursue , and the wage-slave does not have the money food items when he stops working the gruelling, exhausting, and time-consuming minimum wage work to pursue entrepreneurial endeavours. The solitary, demoralizing low-wage job that is unfulfilling and doesn't provide even a quarter of a income to pay for housing, drains the entire sacrosanct, limited time that he could allocate to income-generating assets if his requirements were met in full with a guaranteed basic income.

If everyone had control over their lives and their money by living in a market-based free economy that was totally private and all transactions were voluntary, like how one could earmark the 13 years in their time that people are deprived of because they have to attend the 2,340 days of intoxication K-12 compulsory indoctrination days, wage slavery could be avoided as people would be able to concentrate on creating income-generating assets and building brand awareness as young as. It could lead to them becoming wealthy before they reached the age of. Making income-generating assets to boost your wealth and speed up growth of your brand could take the form of monetizable videos such as informative content, educational podcasts, informative ebooks engaging songs, as well as automated online courses that are the basis of your brand's products offerings.

Achieving a wealth of unimaginable size by creating wealth-generating assets that generate income and building strong brand names via social platforms are attainable even at a young age. This means that you don't need to become a slave in the workplace when you are an adult to purchase food. For instance, it was revealed that "Ryan 7 years old, a content creator, earned $11 million on the revenue he earned from his YouTube channel, landing at the 8th spot on Forbes 2017's list. His YouTube channel has earned him huge numbers of parents and kids who have watched Ryan play with new toys and give his opinions on them" (O'Kane O'Kane, 2018,). Ryan has been very successful in developing a successful brand around his passions that he's made more money in a single month, at the age of seven than the average adult would make from multiple stressful jobs for the entirety of their lives.

If children begin working from an early age, creating brands that reflect their passions and interests they are highly successful when they are adults, particularly when they have already accumulated an impressive amount of experience in the workplace and have developed successful brands during their early years that they are able to further develop during the years of adulthood. Anyone who is successful on the job when they were young children are likely to achieve a significant financial success when they become adults.

After a decade of creating a wealth of income-generating assets and building a solid niche market as a prolific content creator and well-known influencer, the person can invest their savings into purchasing numerous income-generating assets, including bonds or rentals, index funds as well as stocks that have large dividend yields. The smart Millennials know the power of compound interest as well as the long-term financial benefits of living a simple lifestyle to continue to buy ever more income-generating assets until they get to the point where they don't need to be actively working or invest in further building their brand to pay for their expenses when they have already achieved high fame leverage and wealth through the popularity of their own brand content and income-generating assets.

The younger generations are fully aware that attention is the value of attention in the new age. Generation Y is adamant about the importance of entrepreneurship, creation of content and building brand awareness to earn money from selling products or royalties, as well as making advisement revenue. Furthermore, content creators from the generation of millennials recognize the significance of building brand recognition to ensure that they ultimately earn subscription revenues, generate sponsorship income, and earn each month Patreon contributions from their patrons.

Generation Y are conscious of the absence of job opportunities for employees accessible and have realized that they have to build their own jobs without being dependent on any employer to provide an opportunity to work. The jobs of employees are usually ineffective, demoralizing minimum wage, non-fulfilling positions that never provide benefits, pensions or a sustenance pay or even one percent of what your time, effort and effort are worth.

Human workers have become a rare species in the age of digital technology and humans have become obsolete horses in the modern age of automation because they are being replaced by robots, automation, automated procedures as well as artificial intelligence. Degrees are also completely useless in the age of digital technology which is why labor cost has become the highest important expense that employers are trying to eradicate by introducing robots and automated processes, as well as machines as well as artificial intelligence, into workplaces to completely eliminate the requirement for human labor and perform more efficiently.

Robots could be the ideal employee for companies to use because they are cheaper to hire and aren't seen as a threat for the business. Robots are also more efficient and effective in doing specific jobs than humans and don't require pay or pay. Additionally, robots do not require additional benefits "cannot be prone to errors due to having been pre-programmed to perform specific tasks with perfection, will do not make excuses to work, and never be late to work" ("What is an employee? ,").

Robots for industrial manufacturing, kitchen robotics and warehouse robots. Retail robots, robots, hospital robots hotels robots, packing robots semi-automated agricultural machines as well as automated procedures are becoming increasingly widely used on a global scale. For instance, the use of industrial manufacturing robots has had a major impact on manufacturing. The use of industrial robots for manufacturing have not only made manufacturing of goods more feasible at a global level as well as reducing manufacturing flaws and higher quality standards as well as made the cost of products much more affordable. The replacement of human workers in factories to a robot workforce has transformed the industry of manufacturing.

Since almost every task could be reduced to set of specific task, robotics is constantly developing to have the capability to finish the necessary task for increasing numbers of jobs. The utilization of robots by companies will grow more extensive as they become more advanced and capable of doing more than just performing certain tasks, like making meals, transporting luggage and moving warehouse objects and manufacturing items, as well as providing medical assistance, and providing hospitality services. Artificial Intelligence will likely be better able to communicate with humans and absorb information more quickly as machine learning and other technology are developed and more advanced.

Semi-automated and robotic machines can supply customers with instant services without making critical mistakes. In addition, businesses are able to pass on the savings that come with using semi-automated equipment and robots as part of their staff to offer customers cheaper prices, which in turn adds an increase in the value of products and services that are brand owned by the company. Robots' use is not just a matter of "consumers to not just get more bang for their money" ("bang at the") when purchasing items and services as well as be more content with the brand because of the brand's ability surpass their expectations in terms of price, quality of service as well as speed of service and the quality that the services provide.

Robots should be employed in a global manner by businesses to enhance the quality of the services relationships marketing can offer to customers. If the robotics sector was to grow exponentially and become more mainstream, then more markets could be brought to fruition, and a huge number of jobs will be created as the demand for business robots increases. Additionally, the world's gross product could hit an all-time level if the majority of jobs that require basic skills were taken over by robot workers. The use of robots will ultimately enable companies to drastically reduce the cost of labor and deliver services more quickly method. Utilizing robots would enable brands to boost profits, brand growth and brand equity, as well as brand recognition and loyalty.

The customer service experience will also be greatly improved with the help of a robotic workforce. Robots are better able to deal with angry or hostile clients than human counterparts. In addition, robots can be programmed to not display any negative or hostile attitude towards customers unlike human employees who can be driven by emotion. Robots are also able to complete tasks at a much quicker pace than human workers. The elimination of human workforce will enable individuals to become free by freeing their time for satisfying activities instead of giving up on jobs that are unsatisfying in the sense of a symbolic cog of the organization's wheel.

Robots, semi-automated equipment and automated processes may be utilized by companies to improve the quality of their products. For example automatic fast-food service can transmit fast food orders almost instantly to chefs. However an individual fast food service worker may have to write the large order using handwriting that could be very sloppy. They would be required to approach chefs to hand them an nearly unreadable note that outlines the food order. Apart from the use of robots in manufacturing and the restaurant industry, robots are also being used to take over human labor in other fields including the retail industry, hotel industry, and the agricultural industry.

Like humans, the ultimate robotic worker can be duplicated in a large number, making them similar to each other. By buying a number of robots that have a similar design to each other in all aspects, companies are able to consistently offer their customers top quality services. For example there are five or more fast food servers that are automated could be used in fast food restaurants to reduce the need to wait in queues. It's also amazing that robots are programmed to not make mistakes, that is in distinct contrast to human employees who aren't infallible. If employees make errors, it may cause customers to be dissatisfied and damage the image of the company in particular if the customer's expectations weren't satisfied. Fortunately, human mistakes can be prevented by using robots in business globally to replace human workers.

Robots could be the perfect employee for companies to hire because of their incredible capabilities, like their capacity to beat humans in completing tasks with no errors. Robots can also help businesses to offer a greater value to their customers with a much more efficient way that could be passed on to customers through reduced prices for products. The increasing use of machines, robots and computerized software by companies will eventually dramatically increase the gross domestic product because businesses can provide significantly superior quality and services to their customers, especially when their human labor costs dramatically reduce. Although the replacement of humans by robots could result in a net loss in jobs could result in an income that is guaranteed and a significant increase in the quality of life, particularly when the majority of jobs could be outsourced to robotic employees who can effortlessly and constantly finish tasks in a efficient and productive method.

Technological advancements that continue to advance make the world a much more productive and could allow our world civilization to evolve into a type three civilization within 5800 years (Kaku). Technology advancements have led to profoundly beneficial technological advancements that have enabled brands to reduce inefficiencies, increase the efficiency of their workforce as well as reduce operating expenses and boost efficiency.

The increase in technological innovations and advancements could have profound effects for the economy and eventually result in an increase in quality of life and an improvement in living expenses because more business activities are outsourced to robots or automated processes. The continual removal of laborers by robotics and automated processes may be a source of the possibility of a universal guaranteed basic income, as productivity of the economy increases without the requirement for humans, particularly as machine learning allows artificial intelligence to master virtually everything.

Employment as an employee for professionals with grad degrees, doctoral degrees engineering degrees, STEM degrees, nursing degrees law degrees and numerous certificates are now mostly gone with only a handful of jobs open for the vast majority of educated professionals. Employees have gone extinct as a species in the age of automation that is ubiquitous and are viewed as a burden and liabilities for employers who lower their profits.

Humans have become outdated animals in this age of automated processes. They cannot compete with robots, machines artificial intelligence, automated processes, despite the best of their efforts. The absence of a guaranteed income can have devastating consequences as the life expectancy rate continues to plummet into the dark due to people not having enough money to pay for the necessities of life, including housing and food. A lack of a guaranteed income can also affect businesses that offer products and services to their customers because demand decreases for their services and products due to their clients not having access to the lines of credit that can finance the purchases they can't pay for without having income. It's also troubling that the highest-credentialed and educated individuals are unable to get employment as employees, especially since they're burdened with debt and have wasted the last decades of their lives getting degrees that do not possess a single ounce of intrinsic value or even intrinsic value on the market for jobs.

It's a bit disconcerting to learn that an overwhelming majority of STEM graduates are not employed or are doing dead-end jobs at minimum wage. Additionally, it's troubling to know engineers turn to driver for pizza deliveries, physicians are resigned to work as day laborers at farms, nursing assistants find themselves unemployed and lawyers accept working as Uber drivers to earn money in this digital age. If having a college diploma or Doctoral degrees, STEM qualifications law degree, or nursing degrees cannot help to get you a job, it's unquestionably useless in every aspect.

All over the world, STEM degree holders are confronted with an employment crisis, specifically when employers reduce their workforce and move to artificial intelligence and robotics as a substitute for maintaining a human workforce that is outdated. The rate of unemployment among STEM graduates is predicted to increase to an unimaginable number as many STEM degree holders are competing for job opportunities.

"Even although employers might appear to be begging for applicants with science, technology engineering , and math (STEM) capabilities but the truth is that despite the increasing focus upon STEM education, a lot of STEM graduates remain unemployed for six months after graduation in an upcoming CIPD report. This is not just vital for students and universities however, but also for employers. Employers are of the opinion that students must to acquire work-ready skills in addition to their academic studies, so that they can be able to apply their knowledge acquired in the classroom into work ("How employers can be successful," 2018,). Employers' expectations aren't realistic as STEM students are overwhelmed by academics and do not have time to pursue an unpaid internship when they are enrolled at the university. The reality can be that the most skilled STEM graduates are often unemployed or at worst, severely underemployed.

The rate of unemployment for STEM graduates is predicted to increase to an unimaginable number as many STEM degree holders battle for fewer jobs. "All credible research provides similar evidence regarding the STEM workforce: plenty of demand, stagnant wages and, according to industry reports hundreds of candidates for every job advertised. It is important to be concerned about the bleak employment opportunities for the top STEM graduates. In general, U.S. colleges produce more than twice the amount of STEM graduates per year than they can find jobs in these areas" (Salzman 2014). Around the globe, STEM degree holders face an unemployment crisis, in particular since more companies cut back on their staff and shift to artificial intelligence and robotics as a substitute for the outdated human workforce.

A STEM degree could result in unsustainable debt, unemployment or underemployment, as well as in particular, poverty or destitution since the interest of your STEM graduates' student loans is added to the principal every day, which further erodes their prosperity. If they are unemployed or underemployed and underemployed, the STEM graduate could need to wait for several years, if not more in order to pay back their student loan debts if they happen to be fortunate enough of working a non-existent low-paying job that fails to give them a living.

To illustrate, to the disappointment of students who are STEM students with degrees in engineering, it has been discovered that nearly "80 percent" of STEM graduates from India are not employed. The majority of them are required to work in non-STEM fields , or are without a job. One of the primary reason for this is the huge number of engineering schools within India and their shoddy academic infrastructure. The number of engineering colleges in India is 10,396. AICTE recognized engineering institutions The majority of the engineers trained by these colleges are employed. The earlier companies hired engineers, and then taught them how to work with proprietary technologies but this is gradually disappearing off" ("Qualified but not employed," 2018). The employment outlook for STEM graduates is negative and dark, particularly since robots and artificial intelligence will likely to take over workers with higher education in the near future.

Engineers employed by employers could be making themselves vulnerable to future unemployment. "By insisting on automation, engineers could be slowly shedding of their jobs. Artificial intelligence in conjunction with various other technological advancements, has always increased productivity. The more work is completed which means businesses earn more. Scheduling, for example, is one of the tasks that many companies have already automated. The need for automation has led to a variety of new software developments. Additionally, methodologies for projects like Building Information Modelling, place automation at the top of the list" ("Will architects and" 2019.). Similar to other fields of work it is likely that the STEM field will be less populated with employment opportunities in the near future as robots and artificial intelligence start to dominate the outdated human workforce.

The direction of the employment marketplace for engineering professionals is shockingly miserable. "The unemployment rate has been increasing in rate in the majority of countries that are considered to be developed. However, in recent past, it has been observed that STEM graduates with engineering degrees are having to wait some time to find the ideal job in Canada. In the case of experts, it is discovered that jobs that specifically related to engineering are becoming increasingly difficult to get. Based on a study by a organization called Statistics Canada, it is discovered that two out of three engineers work in jobs that are not the traditional engineering. The study was conducted in Ontario and has the largest engineering workforces in Canada. Engineering jobs in the traditional sense are in decline" ("What do you think is happening in the" 2015.). It's a costly error to take on a mountain of debt and not work for three or four years to pursue of a STEM undergraduate degree that is not able to provide any intrinsic worth since engineers are more often in a state of unemployment or are severely underemployed. This issue of engineer unemployment is predicted to intensify overtime particularly since the demand for engineers decreases.

Universities are producing engineers who are not able to find jobs in the decade of

2010 and beyond. STEM degree holders will face higher unemployment rates in the near future because of the rise of robots and artificial intelligence which are able to outperform their human counterparts in every aspect. The root of the problem with STEM graduates not finding work is the fact that they're burdened with close to $100,000 to $200,000 of debt. They've given up working in their youth only to be left with nothing but exorbitant debt, pain and despair. Also that they would be better off not attending college, university, or graduate school because the result of their perseverance was not what they expected and now has caused them to fall into massive amounts of debt they can't get out of even if they had a STEM job.

STEM graduates aren't likely to be able to recover from the loss of not being able to enter the workforce. STEM graduates have been able to avoid creating a huge negative net worth as a result of being burdened by more than a quarter 1 million dollars of debts from attending universities or colleges and graduate school. "Factoring in lost opportunities and the total expense of going to" ("In an alleged," 2018) university to be an STEM graduate is around $200,000. Graduates of STEM who are highly qualified are underemployed, and it's disconcerting that the highest career path that STEM graduates are able to get out of college as employees is the possibility of a dead-end and a minimum wage job that isn't even able to provide them with the food they need.

In addition, in the current modern age, the chances of a lucrative job even after having earned numerous advanced degrees and certificates is bleak and gloomy for students who aren't well-qualified. The reason that leads graduates who are overqualified to come to an impasse in finding lucrative jobs within their specific area of study is employers are only seeking applicants with prior experiences in the specific task they're hiring them to do, which results in an unbreakable barriers to the entry. Even with impressive academic qualifications, graduates are likely to be totally unemployed or extremely underemployed after graduating from university with advanced degrees.

A master's degree, doctoral and STEM degrees do not warrant the high cost of opportunities particularly when chances of securing a job are poor in this digital age. The graduates' disappointment is that the chances of landing an excellent job are probably not much higher than they were before they received their degrees. The problem is that time is the only resource that is not replenished, and your most stable job could be one that you make yourself. You are always working to improve when you're self-employed like a full-time freelancer blogger, live streamer or creator of video content.

The goal of the degree of a university is less valuable than the paper that it's printed on. A STEM degree is for instance, lacking any merit, and is not of intrinsic or extrinsic worth. A document from an institution is something that employers do not consider valuable regardless of the amount of effort put into it to get it. The reality is that pursuing the STEM degree can be an ideal recipe for long-term unemployment, pain and unsustainable debt because the issues that arise of obtaining the STEM degree are a constant in every aspect of your life for a long time as a the lack of financial stability could quickly ruin your life.

Additionally, medical colleges are turning graduates who are unable to find jobs in the decade of 2010 and many medical school graduates are likely to have higher rates of unemployment in the next few years since they will be replaced by AI doctors and medical robots which can beat them in all aspects. Much to the dismay recent graduates of medical schools It was revealed that "nineteen percent of specialists who took their exam for certification in 2017 said they were not able to get immediate job and neurosurgeons, radiation oncologists , and orthopedic surgeons being the most likely to find themselves unemployed. The doctors who wish to practice are unable to get their first step in the door, even those who are working are trying to find operating room time" (Grant 2019). Additionally, "a stunning

78 per percent of the ear nasal and throat (ENT) professionals who completed their training in 2014 were unable to secure a job, and 30 more are set to go on the market" (Blackwell 2015). The medical doctor job shortages are predicted to increase over time, particularly because artificial intelligence systems for medical use and robots are brought into clinics for medical use.

The rate of unemployment for surgeons in the next decade could "reach 75%, which eventually means that the million-dollar medical degree to not even be worth the paper they're made of" (Koetsier (2018)). In the next few years, "whole-body scanners will be equipped to scan our bodies at the cellular levels just entering the clinic. DNA will be automatically analysed and personalized medicines will be developed in real-time using medical 3D printers aided through healthcare AI systems. Necklaces, bracelets or internal sensors will be able to monitor the health, improvement or problems and constantly rely on AI medical professionals and add to our medical knowledge base worldwide knowledge , and improving every diagnosis and treatment constantly and in real-time. When surgery is required advanced medical robots will in charge of everything instantly and on-site" (Koetsier 2018). Similar to nursing schools medical colleges are pumping graduates who were unable to find work in the decade of 2010. It's only natural that graduate from medical schools will be facing increasing unemployment in the near future since they are being replaced by AI doctors and medical robots which can beat their counterparts in all aspects.

The main reason for medical school graduates who are unemployed is the fact that they're saddled with close to $400,000 in debt. They've given up on working in their early years and a portion of their adulthood year only to be left with nothing more than suffering, pain and a sense of discontent. Also it would be more successful not attending universities, colleges, or medical school because the result of their sacrifices was disappointing and now has caused them to be saddled with an enormous amount of debt they are unable to get out of, even without a medical doctor job. "Experts believe that the rollercoaster experience of falling, and then rapidly increasing enrollment in medical schools doctors' unemployment, and currently, medical graduates who are stranded is a result of a variety of causes. For instance, 178 orthopedic surgeons who have been fully trained in Canada are unemployed at present according to the Canadian Orthopedic Association. In addition, waiting times in the field aren't not significant. The tightening of the job market is exacerbated by a increasing number of physicians waiting to retire, and still clinging at their privileges as operating rooms" (Blackwell 2015).

Medical school graduates aren't likely to ever recover from the chance to not be able to enter the workforce. The graduates of the medical schools have fallen victim to the pressure of establishing a hefty negative net worth as a result of being burdened by close to one million dollars worth of due to the fact that they attended the university or college and then medical school. " It is the belief of many that a graduate of medical school will be able to practice as doctors, and that is the case without exclusion. A lot of people are shocked that there are a lot of doctors who aren't able be a doctor. There are many doctors who graduated from medical school with good standing are exiled from the profession they were born to heal, even though they've completed at minimum eight years of advanced training', according to Lynn who is the co-founder of Doctors without Jobs. When you consider lost opportunities and the total price of medical school can be as high as $800,000 as per the website, Best Medical Degrees" ("In the purported year of" 2018.). Medical school graduates who are qualified are extremely unemployed, and it's a pity that the most lucrative job opportunity for those who are medical school graduates are able to achieve after leaving medical school as employees is an unfulfilling and a minimum wage job that does not even offer an income that is sufficient to sustain them.

The process of obtaining the qualifications to become a medical professional after graduating from high school is a difficult undertaking that can take seven or eight years as you'll need to obtain an undergraduate degree as well as be able to pass the MCAT as a prerequisite in order to be accepted into the three-to-four year medical school. If you're determined to become medical doctor, it will enable you to lessen the burden of debt by doing everything within your power to reduce the cost of tuition while you earn your undergraduate degree while studying at the university or college of your choice.

Like physicians, nurses also have to accept a in a solitary, minimum wage job that isn't even able to supply them with food as their educational credentials have been rendered useless in this digital age. Although becoming a registered user to be a feasible job opportunity in the past but obtaining a nursing degree has no benefit. "Nurses additionally have a long-standing tradition of being required to defend their place in the health system. When cheaper health care providers been threatening their existence, now automated technology is doing. Intense nursing care tasks are being handed over to machines. In Japan for example the growing demand for care for the elderly has led to the development of robots for nursing like "Robear". While they do not yet take care of patients, they could become nurses' replacement in the near future. Robots have also entered North American health care. The experts in this field see a need to use them in removing doctors and nurses from monotonous tasks, like bringing medical supplies and transporting food and medicines" (Gionet 2017, 2017).

Nursing degree holders as well as other holders of advanced degrees also are concerned about their main education choices. It's a bit disconcerting to learn that an overwhelming majority of nursing graduates are either unemployed or are terribly underemployed, working low-paying, dead-end job that does not give them enough money to affordable housing.

A shocking number of nurses who have completed their training face an extreme unemployment problem on the job market for nurses that is why they aren't offered jobs by hospitals. "Retirement-age nurses are staying to the end as they require the money. In addition there are many nurses who were on hiatus when they were in their 30s to have children are not doing as they did because their spouses been laid off from their jobs" (Sellers 2013). Nursing degrees are not able to provide some merit and the desire to earn one could leave you unemployed and a lot of debt.

Much to the dismay of nursing graduates, "only 55%-59% of newly graduated nurses with bachelor's degrees were employed during the four years of study (2010-2013). For associate-degree nurses, the number varied between 42% and 45 percent" ("Nursing graduates"college," 2014). Nursing graduates are still unsure about their post secondary education choices, which left them in debt and unemployed. Nursing schools are producing nurses who are unable to find work in the decade of

2010 and beyond. nursing graduates are likely to face higher unemployment in the next years, as they're replaced by AI doctors and medical robots that are able to outperform nurses in all aspects.

The goal of becoming an nurse is unquestionably not justify the expense because it will not only cause you to lose three to ten years of work and incur a significant amount of debt of more than $50,000 however, it also makes you unemployed in the same way as those who have completed medical school who are unsure regarding their post-secondary educational choices. A nursing degree doesn't justify the cost in any way and the quest for a career as a nurse has turned into an unaffordable expense for those who have recently graduated as nurses. It is because the nursing degree they earned has no intrinsic value , or even intrinsic value to employers, which makes them useless.

Like nurses and medical school graduates Law school graduates also face low chances of finding work. The law school curriculum does not justify the high expense of earning the law degree even if it is able to pay for no tuition at the side of the student. It is because of the massive chance cost of sacrificing at least six or seven years of job market after having completed high school, and not finding a job that is guaranteed to open for you in the field after the law degree is obtained is something that cannot be redeemed.

The process of obtaining the qualifications to become a lawyer after graduation from high school is a difficult seven-year process since you must complete an undergraduate degree and take the LSAT in order to be a prerequisite in order to be accepted to a three year law school. The desire to become an attorney is something students who are interested in becoming lawyers should avoid to pursue since it is all the time preordained to result in you being sunk into debt and possibly unemployed or in a glaringly underemployed position after graduation from law school in this digital time. In this digital age attorneys are just a dime dozen because the lawyer job market is extremely saturated.

In this digital age the law school graduates are also a dime a dozen as the job market for lawyers is extremely overcrowded. Much to the dismay students who have recently graduated from law schools, it was revealed that the law school class in 2018 of "Inter American University Puerto Rico had an exorbitant unemployment rate of 32.85 percent, while the law school class of 2018 from Pontifical Catholic University of Puerto Rico" (Zaretsky 2019) was not much with a hefty percentage of unemployment 29.20 percent (Zaretsky 2019,). It was disappointing that other law schools also reported that their law classes in 2018 had low rates of unemployment ranging between "21.53%-27.81 percent. University of San Francisco for instance, reported a shocking 27.81 percentage of unemployment for their law students of 2018. Meanwhile, Mississippi College, Concordia Law along with School University of La Verne" (Zaretsky 2019) had a grim rate of unemployment of nearly 25 percent for their law students of 2018. (Zaretsky 2019). Law school graduates with a law degree are in a very low employment rate and it's extremely disappointing that the most lucrative job opportunity the most highly educated law school graduates will achieve after they graduate the law schools as an employee is an unfulfilling with a minimum wage job that isn't even able to give them a source of income.

It is logical that the jobs of workers have gone out of fashion and jobs from the past decade that were once provided to graduates with high-level credentials like doctors, lawyers, nurses, engineers and doctors are no longer available in the modern time. In the modern time, the chance of securing a job is yours after years of study is about as low as the chance that you will win the lottery. The handful of professionals who are employed in private sector jobs where the demand for their services is free will experience similar outcomes to their non-employed counterparts. The professionals currently employed will scheduled to be dismissed out of the workforce when they become displaced by artificial intelligence, robots machines, automated processes that are able to surpass them in all aspects.

The rise in machine-learning has made human beings apathetic employees and outdated horses in the age of automation, as artificial intelligence and robots are able to surpass humans when it comes to completing any set of repetitive tasks lacking the ability to think creatively. Human workers are for away from being a thing of the past human beings have turned into obsolete horse breeds in the time automation because of being replaced by a robot workforce.

Jobs that are likely to be available in the near future will likely be held by entrepreneurs, content creators performers, musicians authors, artists comedy directors, movie directors and performers, educators from around the world and influencers who must continually innovate and harness their creative abilities to create greater intellectual wealth online. It will be harder for artificial intelligence to do tasks in the near future that require a lot of imagination. It is likely that artificial intelligence can create intellectual works as intricate and dynamic, exciting and innovative such as Star Wars, Harry Potter and Lord of The Rings. Lord of The Rings anytime in the near future.

Much to the dismay of the person who is interested The various opportunities for employment that remain open to them as an employee can be disappointing, unfulfilling and unfulfilling minimum wage terrible, unwelcome, and incredibly dreadful jobs that provide even a fraction of a livable amount to pay for housing. The surrender of your finite, sacrosanct irreplaceable, essential, non-replenishable currency for a meager fixed amount of currency in fiat that is provided on a predetermined basis in exchange for a painful, dead-end in a time-consuming, non-fulfilling minimum wage job is an opportunity to plunge you into a world filled with wage inequity and extreme poverty.

The reason is that the positions offered by these employers aren't paying you at all a fraction of a food and a decent wage to afford housing, and force you to support yourself through credit card debt that increases your net worth that is a huge negative and also increases the amount of debt you have. The employee positions don't offer any advantages or pensions, and they deplete the time the worker will require to build income-generating assets to get him out of in a life of struggle, financial hardships, wage slavery with a plethora of problems, exhausting labors and insurmountable debts.

To the chagrin of the wage slaves, the employers determine the time they can eat and use the bathroom and offer wage slaves the slightest benefit for their labor and effort. This type of life of constantly working while battling constant loss of wealth causes chronic diseases that cause DNA damage as well as telomere lengthening and anxiety, depression and an increase in mortality as the wage slave struggles to face the gruelling realities of his miserable life , as he continuously kneels before his employer without any basic guarantee of income that can provide an insurance policy in order to avoid the squalor of wage slavery.

As previously mentioned that if someone is eviscerated for ever, deprived of their lives, time liberty, wealth and happiness, it is the conditions for unbreakable wage slavery, particularly in the absence of a guaranteed income that can provide the wage-slave with an escape from the precarious position by having enough money an entrepreneurial career that is self-employed. The wage-slave will require immediate relief with the biweekly or weekly paycheck just to pay for food and doesn't have the privilege of being self-employed while being eviscerated for the rest of his life out of his time, his life liberty, wealth and happiness despite his fact that will not make the minimum of a quarter of a living wage to afford housing.

In a market-based system of control, you are not in control of the way your hard-earned money is used. For example aside from there being more than 13,000 evisceration charges that are imposed by bureaucratic systems against an person, the individual is also required to pay for insurance on their car as well as the registration fee, as well as the fee for renewing their driver's license at the end of each year just to use their personal vehicle.

For instance, in New York City for instance there is a high likelihood that you'll require a permit soon to be able to use a bicycle, and it shouldn't be a surprise that you have to pay the cost of registration for a bicycle and the fee to renew your bicycle license at the end of time just to take your bicycle out on the street. Due to the structure of this market system that is controlled that the person has little to no control over how their hard-earned money is spent , which is an environment of wages slavery because the individual is required to earn more money to pay for their ever increasing cost demands that are artificially created in a controlled market like the mandatory obligation to purchase auto insurance to be permitted to drive your private vehicle to work, and also pay a high registration fee each year. In a free market system it will be simpler for a person to climb out of the filth and avoid a life of slavery to wages.

One of the primary causes of extreme poverty which is difficult to overcome is that it involves adversity to financial challenges and issues which make life more difficult in all aspects because money is like oxygen in a market-based economy that is controlled and therefore is required to cover all aspects of life, even getting the necessities that are relegated to the lowest tiers in Maslow's Hierarchy of Needs pyramid. Lack of purchasing power that prevents them from being in a position to meet their basic necessities, like food and shelter and causes them to fall into in extreme poverty. They then turn to working low-paying, dead-end wage job as a wage slave to gain the money to purchase some food items. A dead-end, minimal wage employment as an employee can keep the wage slave trapped in extreme poverty, and suffering from wage slavery because it doesn't offer even a fraction of a daily wage, it also takes away the time to generate income to have the possibility of getting out of wage slavery.

The issue of getting out of wage slavery poses a problem that is attainable to overcome , even if it demands the wage worker to make some drastic concessions in order in order to escape the shackles of an extremely depressing exhausting, time-consuming, demoralizing and dead-end, minimum wage job with little in the way of benefits, nor any sort of satisfaction. The wage-slave must realize that they will have to spend for at least 100 hours a week to turn this dream and dream of ending the slavery of wage work into reality.

To get away from the squalor of slavery to wages is just an exercise in numbers. That means you have to take every step to reduce the cost of living to ensure that you're not dependent on a wage per hour to be able to survive. The methods you can cut down on your expenditure are limitless. However, the main ways to cut down on your expenses include living with friends or family members or carpooling with other people and eating only at home, not ordering from restaurants and avoiding spending money on frills. Also you might need to be a minimalist to cut down on your expenses for the month.

Once you've reduced your costs as much as practical, you'll have to dedicate the time you can afford to put into building your brand and making as many income-generating assets as is possible including podcasts, blogs, videos and audiobooks, eBooks, articles, and online courses. That is you'll need to produce by monetizing as many income-generating assets as you can so that you are able to wean yourself from giving up your precious, sacred indispensible, irresistible time in exchange for the fiat currency which is granted to you after you have completed in your miserable, unending minimal wage job. The wage slave with the greatest desire to get out of wage slavery might accept working more than 120 hours a week with only a few hours of sleep to generate additional income-generating assets with the possibility of getting out of wage slavery.

When it comes to cost of living that is rising and the rising cost of living, the wage slave's dollars will not stretch very far as the minuscule amount is virtually nonexistent purchasing power in this age. In 1913, the value of $10 adjusted for inflation would be approximately $990 today. Also that dollars have lost 90% of its buying capacity since 1913 and has become practically useless in today's digital age. If you take into account sales taxes, a dollar won't purchase the smallest item at stores that sell dollars.

In an economy that is controlled where the hard-earned money is used to pay for an trust account, as well as amenities like vacations, pensions and vacations. luxury, benefits, and the exorbitant wages of the public's glorified welfare aristocrats and at the expense of the person's minimum wages and time, as well as wealth as well as dignity, happiness and health. The person is able to destroy his health by working for a exhausting, time-consuming disconcerting, depressing, and dead end and a minimum wage that takes away his time to sleep well.

A gruelling and exhausting, dismal interminable, low-paying wage job entails the life of struggle as well as suffering, pain as well as destitution since the wage slaves in poverty can't even afford housing and have enough time to have enough sleeping. The lifestyle of a wage slave which the evisceration charges enforced by bureaucratic structures drain the workers out of a substantial amount of their minimal wage and time, resulting in perpetual stress, fatigue and perpetual poverty as they are not afforded the luxury of being able afford to sleepor eat, or even be in control of their own schedule , which prevents their having the time required to build income-generating assets with the possibility of getting out of the slavery of wages.

In California the state of California, a couple earning an income of $100,000 trying to reside within San Francisco for instance is classified as a family with a low income because the cost of living is extravagantly expensive. If you take into account all the fees for evisceration that bureaucratic structures impose such as auto insurance rates as well as car registration fees car oil changes as well as maintenance and repair expenses and internet subscription fees as well as the expense of grocery shopping and utility bills, the cost of gas, the expense of renting, the cost of accruing interest charges on credit card debt and the cost of accumulating student loan debt, you will comprehend why the majority of people will have at least $67,000 in debt, which is their net worth after they die. That is the cost of living is too expensive and forces the person to accept the role of an employee slave to have enough money to purchase food items and pay the ever-increasing expenses that continue to increase the burden of debt they carry.

Living expenses are expensive, particularly when sharing an apartment in a shoebox with multiple roommates consume at the very least 2/3 of your total income. The cost of goods and services continue to rise as the buying power of wages continues to decline each day, and eventually leads to an environment of perpetual wage slavery.

Wage slaves live through debt financing that can take the type of payday loans as well as credit card debt mortgages and other types of debt they use to fund the necessities of survival. They usually pay bank charges and a 20-28 percentage interest rate on credit card debt, which the interest accrues into the principal every day. The burden of debt increases each day with the price of living, as their infinisimal minimum wages decrease in purchasing power each day due to inflation. The wage slave lives an arduous life of debt and can't pay for the expenses of living with their inexpensive minimum wage, which isn't enough to cover all their essential requirements for survival. This is why nearly everything they purchase is funded by debt finance. The wage slaves are highly dependent on debt because the price of living has forced their market share and unable to afford essential necessities for survival. With the price of living steadily increasing wages, wage slaves is not able to pay the bills in order to secure an income-producing dead end job, as they constantly require money to cover their growing costs. Food and housing aren't free for the wage slave , and are becoming increasingly cost prohibitive despite new technological advances that have not brought cost of these essential necessities down.

Being in a soul-sucking exhausting, draining, demoralizing interminable working for a minimum wage as wage slaves and keeping yourself from debt financing is an ideal situation for you to be stuck in extreme poverty while your debt load increases as your buying power diminishes from your pathetic minimum wage that never keeps pace with inflation rate, leading you to lose increasing purchasing power each year. The jobs of employees are awfully miserable and heart crushing because they are so disappointing, inexplicably abysmal in their meaning, are void of any significance, and are low in pay that they're only undergoing to ensure to ensure that the wage slave can at the very least afford to purchase some food on his minimum salary.

As we've said before, even though being unable to negotiate is not ideal, employers recognize the fact that their employees can be a dime dozen and see them as a petty, capital animals that can continuously spend thousands of hours each year to earn money. What is the reason for wage slavery other than who lack any negotiating ability due to the need for money to eat immediately is that the jobs that are offered to them are very energy draining, soul crushing disconcerting, demoralizing, dead-end jobs at minimum wage that don't provide enough income to enable those who are slaves to cover their expenses in full, pay off debts, or have enough funds to invest and save. The extremely heart-breaking dismal, draining, and demoralizing, and dead-end minimum wage job offers an individual the chance to purchase food at the supermarket for food if they have to work the gruelling hours of more than 40 hours each week. When a wage-slave is not able to perform his duties, he will not be paid a dime by the business for the time it was not worked unless he already has earned time-off days that require thousands of hours to accumulate and end up being wiped out after they have been earned. However, for those who are wage workers, it's extremely exhausting and exhausting to go through all that work to even get a pay day off.

The primary causes of wage slavery are many and include jobs that aren't that pay a living income, university degrees, and professional degrees not being of any value as well as the ever-increasing cost of living making the cost of survival even more prohibitive every year. The individual's wealth when they've reneged on having to work at least 2340 days or 32,760 hours in order to create their brand and income-generating assets in their youth due to having attended an intoxicating 13-year-old school for compulsory indoctrination and completing thousands of unimportant homework assignments, which drained the students of many hours of precious, valuable time, despite their own choice. Because they didn't build the brands or the income-generating assets they had during their youth, they've no money coming in during adulthood to buy food items. This also reduces the person's wealth as over 13,000 evisceration costs continue to strip them of their hard-earned cash. Anyone who is employed in an unfulfilling low-wage private sector job pays for the facilities, pensions and benefits of other people at the expense of their own time, wealth dignity, health and well-being, even while their miserable frustrating, dismal, unfulfilling terrible minimum wage job does not provide enough money to cover the many needs of life, like housing.

Although it can be difficult, it is possible to get yourself from the pile of mud in your adulthood and escape the financial pitfalls of in a world of wage slavery that sees you get more and more indebted each year and are at a dead end minimal wage, a dispiriting ineffective, unfulfilling, extremely gruelling job which pays you an incredibly low amount of income which decreases your purchasing power each year as the increase in inflation. Anyone who is able to make use of their skills to create immense wealth online are aware that the formula for generating wealth.

Contrary to what you imagine the formula for creating wealth is not complicated and can be used to determine the future prospects of profitability for a business. The formula for creating wealth includes the following factors: scalability plus the size of the business plus marketing efficiency = wealth. This formula could be used as an indicator that can not only determine the future profit potential of an online business as well as its long-term viability. Also, if the online business is severely not performing well in any of the factors, for instance, the capacity, scale or marketing efficiency the future of that business will be in danger and be accompanied by grave issues, particularly as the competition within their market continues to be growing.

Making huge profits through social media platforms requires you to work tirelessly all week long to increase your company's content pool by consistently creating a plethora of income-generating assets. It is also necessary to consistently create original engaging content that provides an unimaginable value to your subject matter if you wish to stand a good chances of making huge profits.

If you're a creator of content, you'll also be given the chance to create videos and upload them to video hosting websites, like YouTube as well as DailyMotion. In addition, you are able to broadcast live video content on live stream platforms like Twitch or Smashcast.TV. Live streamers and creators of content make money from donations, advertising revenue, subscription revenue royalty, affiliate marketing revenue and sponsorship revenues. Viewers can donate to content creators through platforms like Patreon, Muxy, and Streamlabs as well as purchase their merchandise with their logos through the online store of the creator like Teespring Store. Teespring Store.

If you're writing, you may post your blog posts on blogs such as Blogger and Medium. If you're an author, you are able to publish your eBooks using Amazon Kindle Direct Publishing and you can also upload your audio books on Audible with the hope of getting royalties from sales.

If you're a programmer, then you are able to create your own mobile apps and games. Mobile games and mobile apps can be made available to platforms like Google Play Store, the Apple Store, and the Amazon Appstore, and can enable royalty payouts from the sales of your product.

If you're a musician then you are able to upload your own music to music streaming services, like Spotify and Pandora and earn royalties from your own music that are played on these streaming services. The purchase of a computer and an internet connectivity will provide you with the chances to be an influencer, expand your brand while working remotely and gain a huge amount of fame.

Maximizing your earnings is possible through achieving high fame leverage and establishing high-paying brand names on the social platforms. Although it could require thousand of hours strategic and relentlessly working to build your brand to the point where they are profitable but the truth of the scenario is that the opportunities to generate income-generating assets, create online brands remote and even become an online influencer is not accessible without a computer and an internet access. Also, purchasing the best computer available and the cost of broadband internet could prove to be the most lucrative long-term investment that you've made in your entire life when you can gain fame and fame leverage from the increasing the number of businesses on platforms on social media, as an influencer within your field of specialization. A person who has achieved high fame leverage could earn much more money in just a day than the average individual is able to earn by working numerous mundane jobs throughout their career.

Earning a lot of money online via social media platforms requires making a myriad of income-generating assets like viral videos, well-known automated online courses, well-known books, well-known mobile applications as well as popular mobile games and songs that are viral. Assets that generate income can come as a result of live videos, archived streams, songs, audiobooks, podcasts eBooks, physical book printed upon demand, web-based classes and software including mobile games and mobile apps. These assets that generate income allow you to earn wealth in any moment and don't limit the potential of earning an incredibly small minimum wage offered by employers that do not even offer a basic income to afford housing.

When you achieve a high level of fame and popularity on these platforms you'll be able generate a lot of riches from sponsorship revenues as well as affiliate marketing revenue donations, subscription revenue or royalties and sales. After you've earned fame through the social platforms for being an influential or content creator in your field of expertise for example, cooking or fashion Your followers will be more likely to consume what you create and also be more inclined to purchase whatever products you advocate buying.

Gaining a high degree of fame leverage so that your income-generating assets are consumed in a massive way by your intended audience begins as an authority and then marketing your content to increase brand credibility, brand loyalty in addition to brand recognition. Your goal is for your company's social media channels to be the top social media channels for your particular niche. As you build your brand's credibility and following on your social media platforms in the future, you'll be able to charge whatever outrageous price you feel is appropriate to sell the single product that you place on your company's social media accounts like offering $1,000,000 for a single product placement that is in the form of an Instagram image, similar to Kylie Jenner. Brands may be able to justify paying $1,000,000 for a single placement if their social media pages have enough followers in their intended audience. This is the main reason it's more important now than in the past to gain high-profile fame and fame leverage, to allow you to achieve fame and fortune through having your brand be the most sought-after and well-known social platform for your particular specialty content.

The more captivating content you produce the better chance that you'll earn enormous financial wealth online. This is due to the fact that captivating content is likely to go everywhere at any moment and the chances of having a massive success and fame advantage through social media platforms increase when you create more appealing content that falls under your particular niche category of content. Producing a large amount of income-generating assets as well as branding content is of utmost importance for content creators as well as influencers that want to stand the chance to earn huge riches online.

The most effective income-generating resources to produce in a massive way for the purpose of generating massive wealth online are viral videos, well-known ebooks, digital courses and popular mobile apps and games for mobile phones, and popular songs as they have enabled a huge number of people to become multimillionaires. The formula for creating wealth can be used to discover the way these products have made people multimillionaires.

Digital products are able to be scalable in a limitless way. capacity since they are able to be distributed globally through multiple distribution platforms, giving you the possibility of creating money time from any part of your intended market. Digital products be extremely powerful since you are able to charge whatever you want buyers to purchase your digital goods. In addition, as per viral videos, you may use the descriptions of viral videos as advertising real estate to sell products with product places. Certain influencers, creators of content as well as global educators offer students over $2,000 to take an online classes.

The effectiveness of your marketing advertising on social media and the enticing pay-per click digital market campaigns to promote your digital offerings will have huge impact on your capacity to make a huge amount of money. Pay-per-click digital marketing campaigns enable you to reach millions of potential customers who might be inclined to take advantage of your content or purchase your product. The investment you make in growing your social media following via Pay-per-click digital marketing campaigns which do not end up selling products could not be a total wasted marketing budget. It's because it can provide you with more fame and credibility as you build your followers, which is advantageous in selling future product places at higher premium costs to businesses. Companies would prefer buying products from famous influencers that have the capability to reach more people by their target audience than influencers with less followings.

If you have the ability to create large followings across several social media channels , then you are more likely to earn immense wealth, even if you choose to limit the volume of content you publish after you have earned an unmistakable image of not only having huge fame and fame but being the top influencer within your particular field of knowledge.

The ability to achieve extreme fame is the first step to being capable of achieving the highest amount of wealth. You will have to increase the number of viewers and focus on your branded content to be in the position of stimulating sales for your product and earn enormous wealth. Influencers and content creators who do not have their brand-name products viewed by a large number of people or are well recognized will be unable to generate revenues since the attention of the customers should be directed to your brand-name content and digital products in order to stand the chance of earning revenue. Companies, like Amazon and AT&T are even able to deplete the marketing budgets of multibillion dollars each year in order to ensure that attention of customers is directed towards their services and/or products. If you gain fame and by being a well-known influencer or prolific creator of content and then achieving extreme wealth via the consumption of your brand's content and the products you sell becomes much simpler and less arduous.

Although becoming a popular online influencer in the current age may seem like an impossible, long-term job as the competition for social media platforms continues to rise to an unimaginable extent and is now more feasible than ever before. However the tried and tested method for achieving a well-known online profile as an influencer will continue to be a success for those who are determined to follow an incremental process to make themselves influential of their respective fields. Also the process to become an influential influencer is broken down into a variety of steps that begin by identifying an appropriate niche sub-genre that you can create content on that the world can take a look at.

In the initial stage, you should produce content that is subsumed in the niche you love since you'll be writing content for a long time as an influencer, and you don't want your career to turn into one of the most difficult, stressful exhausting, painful, and painful undertaking. It is also important to concentrate on creating content for a sub-genre of your niche that isn't over-saturated by excessive competition. This will ensure that the growth of your reputation as an influencer is less daunting and easier process.

The second step, after you've selected the niche genre the content you are going to be subsumed under, you'll need to think strategically about the way your content will differ from videos of competitors. It is essential to develop your strategy for content and outline the ways you can provide unbeatable value to the market you intend to target. "Part of establishing a strong strategy for content is to provide your customers the appropriate mix of informational material and personal content. One reason people are drawn to influencers is because of their authenticity. Influencers are renowned for being capable of providing valuable information to their followers. They also share articles written by other influencers who they think their followers will enjoy. In addition sharing content written by other influential people in your field will allow you to in gaining their attention slowly. This means it's more easy to reach them and solicit for them to help you in the future" (Payne, n.d.).

When it comes to content strategy there is a vast number of ways an influencer can be valuable as a prolific creator which goes beyond sharing the captivating content of other users on the social networks. For instance, if , for example, you own an YouTube channel where you make videos that are based on to comic books featuring superheroes It isn't enough to simply review comics, talk about news about comic books, or anticipate the way events will play out in the upcoming Marvel comics, or DC Comics storylines. There is also the possibility of having to develop your own comic book theory videos. You could also include videos that explain the tales of alternate plots that you have created for the comics or to make biographical videos of comic book characters to provide insight into the vast backgrounds of certain characters. Furthermore, you can recreate comic book scenes using action figures fighting with the help of stop motion footage. In addition, you can make funny superhero prank videos in public , and give tutorials on how beginner artist can create their most loved comic characters from comic books.

Ideas for creating video content of a specific niche are endless and plentiful. You'll need to be extremely imaginative, creative as well as ingenuous, and continually explore the boundaries of the literal box, if you are looking to become an influencer as you'll require a substantial brand recognition to become an influencer who is worthy of endorsement deals. Making content that is similar to those of your the competition, particularly in saturated markets for user-generated content is a recipe for unrelenting failure.

It is essential to reduce the content you are focusing on to make it obvious to your readers the kind of content your audience can anticipate from you channels. You don't want to cause a flurry of confusion from your fans due to it not being clear what your brand stands for and is built around, nor do you want to fall prey to losing followers by creating content that is a mix of genres that have no connection. "Focusing on a specific topic for all your social media accounts lets you present your followers with an overall image of your identity as a person, what you do and what you believe in. This way, you'll be capable of publishing more diverse content, while keeping your profile clean and consistent. It also helps you categorize different kinds of content that you share. This way, your fans will know where they can access the content they require from you" (Payne, n.d.).

According to 3rd step after you've chosen the sub-genre that is relevant to your business and carefully planned out the best way to distinguish your video content from other competitors videos The next step is to develop social media accounts and then carefully select the social media platforms you'd like to market your content on. Like we said, well-known social media platforms include YouTube, Instagram, FaceBook, Twitch, Twitter, Tumblr, LinkedIn, and Pinterest. It is recommended to concentrate on making use of at least one of these most prominent social platforms to host your content. The platform that you choose to host your content on has the highest amount of visitors from your market like YouTube for those who are also video content creator , or Instagram when you're primarily a creating content using photos.

As mentioned previously, the additional social networks you don't explicitly host content such as Twitter and FaceBook can be used by you as marketing platforms for the purpose of distributing your content widely to increase your reach on the internet. If viewers come across your content via these social media platforms which your content has been posted to via content marketing messages and messages, they'll be able to redirect them to the origin of the content. After they've migrated onto the main channel the contents are hosted on and be able to subscribe for your channels if they're curious about the upcoming content that you are planning to roll out in the near future.

As was previously mentioned, like the independent game creator, the influencer can benefit from using IFTTT to create recipes that make it easier to distribute his content on social media marketing. This is due to the fact that IFTTT can automatically distribute the content of an influencer including their YouTube video content, on multiple platforms at the same time like Twitter, Pinterest, Tumblr, Blogger, and others social media sites. The influencer is able to set IFTTT recipes to activate when there are the upload of new videos to YouTube. Through automating content marketing as well as social media-related marketing strategies through the use of IFTTT recipes, influencers will be able to spend more creating content, and spend spending less time marketing the content.

In four steps, after you've created your IFTTT recipes as well as your social media accounts and your content strategy on how you can add a significant distinctiveness to the body of user-generated content from your genre, you will have to begin making your social media content regularly, which will more than likely to be video-based content as videos are the most popular media. It is essential to make your content appealing distinctive, original and worthy of sharing. It is essential to continuously and consistently create enticing content to increase the chances of your content getting momentum and becoming viral.

For influencers, it's important to produce as many videos as you can to build the brand. Videos are much greater in their likelihood to get watched as well as shared than written content and, consequently, have a higher possibility of being viral than other kinds of content created by users.

Videos that go viral, for example, will significantly assist a brand boost profits, brand growth, branding recognition and loyalty and brand equity , as the impact they can have in bringing a brand's success up to the next stage is significant. It is difficult for a video to be considered a viral in the highly intensely competitive market for video It is crucial that the creators of content focus on making as many videos as they can that have the potential to become viral. These videos must include the essential components of a viral film to greatly increase their chances of being viral.

Videos are usually personal, relatable funny, and genuine. Additionally, videos that are viral contain a large number of back links that bring viewers to them. In addition, viral videos use appealing titles, the most effective keywords to tag them and precise video descriptions. The videos that are viral are extremely engaging and are naturally shared with other viewers by viewers. Influencers are able to include their contact information in the video's description to ensure that media companies, sponsors or fans influencers are able to reach them.

Numerous factors contribute to videos to go viral. It is more than simply using the most attractive thumbnail and the most captivating title that is appealing to the viewers and encourages viewers to take the time to view the video. Additionally, it is not limited to using the best and most relevant long tail keywords as well as short tail keywords to be used for search engine purposes relevant to the video's content. Influencers must be smart about marketing and know how to share their videos to the most different social networks as they can for marketing purposes , including, FaceBook, Twitter, Tumblr and Pineterest to aid their videos in get traction organically.

The prospective viral video to possess an element of wow that makes it so captivating that viewers want to share it organically with family and friends. Also, it must be distinct in the sense that it's not easily duplicated by another creators of content. Furthermore, the viral video must also be a bit bizarre and encourages users to engage with the video, as well as an ongoing conversation in the comments. For instance, many people enjoy watching videos that relate to supernatural phenomena because it's an area that people find fascinating that they love to discuss and read the comments on and want to share their excitement with family and friends.

Another aspect that can have an impact on whether the video is viral is its level of humor. The greater the humour power an video has the greater chance to become viral in the course of time. People love watching pranks and sharing them with relatives and friends. Prank videos don't only hilarious and funny however, they can also be very funny and can provide plenty of laughter to viewers.

Influencers must take advantage of more opportunities to collaborate with other influential influencers in order they can increase their followers. When you create viral video content is a great way to make more revenues, get interview requests by media organizations, and create a large following. The advantages of making video content that is viral are endless to build a brand and becoming an influencer of note. The consequences of making videos that are viral are positive and do not have any negative consequences.

It is important to produce viral videos to increase the visibility of your business. If you create video content with the potential to become viral, you will boost your growth in popularity and the number of subscribers. You'll also get more monthly contributions via Patreon because the increase in viewership levels lead to you receiving more pledges from patrons new patrons. Furthermore, by making videos that are viral and generating a lot of traffic, you'll be more likely to receive additional sponsorship and affiliate marketing revenues. Companies are keen to work with influencers who have achieved immense fame and huge numbers of followers, especially when the influencer's followers are included in the targeted market. The most popular viral videos have a long-lasting impact and can be a way for influencers to gain significantly greater brand exposure over the course of time.

In the five-step process, you'll have to systematically, effectively and effectively market your content that is branded to grow your brand's popularity and increase your visibility as an influencer building an immense brand. Like we said earlier, similar to the video game creator the rising influencer is capable of effectively leverage social media platforms to serve marketing and launch low-cost pay-per-click campaigns using low-cost advertising platforms like FaceBook and Instagram in order to not only redirect their marketing efforts but also to gain a more targeted exposure in a cost-effective way than traditional strategies for marketing. Influencers who are growing can purchase endorsements from famous influencers that his audience follows with the hope of rapidly increasing his followers and reaching out to a larger number of people.

Like we said earlier, similar to the developer of video games the rising influencer could increase his marketing efforts in order to not only bring in an increase in viewership and increase the number of viewers, but also to ensure that he's maximizing his marketing budget to the maximum extent feasible. The influencer needs to make sure that his pay-per-click advertising campaigns are targeting the correct audience segment that is subsumed within his target market.

As was previously mentioned in the previous paragraph, the statistical data derived from their marketing campaigns need to be carefully analyzed to ensure that the marketer can not only tweak the marketing campaign as required by removing elements that aren't profitable and ensuring that they also know the types of ads that result in the highest conversions. It is possible to optimize his marketing efforts online to be able to increase amount of viewers to subscribers per dollar.

As we've mentioned before, although making attractive pay-per-click digital marketing campaigns are beneficial for boosting branding however, the influencer must focus on organically promoting his content in order to not fall victim to an unfavorable return on investment. Although it can be helpful to boost brand awareness but it is not guaranteed that investing in attractive pay-per-click digital marketing campaigns or buying endorsements from famous influencers will yield an eventual positive return on investment, even if it helps your business attract thousands of new customers. The most cautious influencer with no marketing budget should benefit from a shrewd marketing their brand "on the social networks like Facebook, Twitter, Instagram, Tumblr, and YouTube for free or at minimal cost" (Totka 2017) which could help bring the attention of more customers to your brand in a highly globally competitive market.

Being a hugely influential influencer through social media platforms in the age of the internet isn't an simple task, it's an realistic goal that is feasible. The achievement of becoming a highly influential influencer via social platforms feasible when you are able to consistently create engaging and valuable content that is categorized within their specific content category and is also achievable for those who have the ability to display unwavering determination, unrelenting determination, and a profound level of marketing efficiency. Being a highly influential influencer through social media platforms can be an extremely difficult task that could require thousands of hours or years of undeniable effort and determination for the most committed and dedicated people.

The numerous benefits to being a successful influencer on the internet and achieving a high fame and fame-building can greatly enhance the quality of your life in every aspect particularly when making cash is not a difficult process, but instead an effortless procedure. If you are able to earn the riches that come with being an influential influencer and influencer, you can free up your time and concentrate on what is truly important in life , not being held back by the work ethic that is exhausting, time-consuming and demoralizing and minimum wage which doesn't even provide even a semblance of a pay.

As previously mentioned, those who have achieved a high level of fame leverage, even if they aren't regarded as top industry experts, can earn more money in the span of a single hour than the average individual make throughout the course of their lives. For example, "at the 2018 10X Growth Conference held by Grant Cardone and Russell Brunson the co-founders of Clickfunnels have been successful in selling $3,2000 within 90 minutes on the stage" (Nanor, 2018,). Influencers with huge followings, who have no formal training are able to earn more in just a few minutes posting one post on Instagram more than the average person could earn for the duration of their lives. For example, "Kylie Jenner commands $1,000,000 per Instagram post, and Selena Gomez next with $800,000 for each post" (Vanderberg in 2018,). The importance of creating an established brand that establishes yourself as a content producer and an influencer in order that you can build an impressive followers on various social networks is usually neglected in comparison to a traditional unfulfilling job that offers no lucrative career prospects.

In the past by the media, it was said that "Ryan was a seven-year-old content creator, earned $11 million through his YouTube channel, landing at 8th place on Forbes' list of 2017's top performers. His YouTube channel has earned him an enormous number of parents and kids who enjoy watching Ryan play with new toys and then share his thoughts about them" (O'Kane O'Kane, 2018,). He's actually extremely successful in developing a successful brand around his interests that he's made more money in a single month at the age of 7 than an typical STEM degree holder could ever make from many jobs over the course of their lives. When people start working from an early age on creating businesses around their passions and interests they are highly successful when they reach adulthood, especially those who have an extensive amount of entrepreneurial experience and already established successful brands in their early years that they will further expand in the years of adulthood. The advantages of becoming an influencer during your childhood can help you to make a huge fortune when you are a young adult. Ryan even has a complete collection of toys named after his life of fun, called Ryan's World, that can be found in the largest retailers, like Target as well as Walmart.

As I mentioned earlier in the past, attention is definitely the value of money in the new age. Generation Y is fully aware of the significance of entrepreneurship as well as building brand to earn wealth through selling products or royalties, as well as making advisement revenue. Additionally, the millennial generation of influencers and creators of content recognize the significance of a building a brand to ensure that they ultimately earn subscription revenues, generate sponsorship income, and earn each month Patreon contributions from their patrons. People who succeed in real life as young people in the social media world are destined to be financially successful in adulthood.

If you can become a successful influencer on the internet and achieving the highest fame leverage, you'll be able to free up your time, not have to work to earn money, and concentrate on achieving your goals and improving your health as well as helping other people. The benefits of being a successful influencer on the internet and achieving a high level of fame leverage confers on the famous influencers are endless as there are infinite possibilities of what you can do to improve your career when you control the entire course of your time, and possess the resources to make your desires into reality.

The many advantages of becoming a successful influencer on the internet and achieving extreme fame through social media platforms is immense. In the first place as an influencer, you'll be able to significantly increase your income and earning potential. Influencers don't have to limit their earning potential being restricted by a tiny set amount in fiat currencies, from a minimum wage, dead-end job that isn't even able to provide a steady income for the cost of housing. Influencers are able to earn earnings from affiliate marketing revenue as well as sponsorship marketing revenue royalty revenue, subscriptions as well as advertisement revenue and donations. Influencers earn money each time donations are received through Muxy, Streamlabs, Twitch cheers or Patreon.

For instance, the Live streamer and influencer Ice Poseidon was able to make over $5000 within the span of 8 hours through donations via live streaming while in bed. In the live stream, viewers would continue to submit 100+ dollars to try to get him to wake up through donations from media sharing and donations via text to speech play on Streamlabs. Donations help shape the future content of the streamer or be able to interact with the broadcaster or get their opinions about the subject.

In addition, influencers earn income from affiliate marketing by referring their followers to products that they can purchase and earning the commission for each sale that is purely profitable for the influencer's. Influencers and content creators for instance , can sign up to affiliate programs and encourage their followers to purchase products through the use of their Amazon Affiliate links (Stephenson 2019.) as well as eBay affiliate links to bring more passive income streams.

Influencers can make a huge amount of monthly revenue through YouTube as well as Twitch through their paying subscribers. They may also generate a large amount of advertising income from viewers who view their commercialized content. Additionally, they can charge hefty fees for product placement which can range from many thousands to feature products of companies in their videos. Influencers also can earn millions of dollars of sponsorship revenues by promoting the brand. Rodger Federer, for example, earned over $86 million in endorsement income (Badenhausen 2019, 2019) in the year 2019 alone. "In along with his blockbuster Uniqlo agreement, Federer added a multimillion-dollar agreement with Rimowa in the year prior. The luggage company joined Credit Suisse, Mercedes-Benz, Rolex, Moet & Chandon, Barilla and others in Federer's endorsement stable" (Badenhausen 2019).

In addition, by becoming an influencer, you'll get an enormous amount of merchandise and access to a variety of subscription-based services for no cost. Brands are eager to sponsor influencers in order to increase awareness of their products and services. Influencers can save money through availing themselves of the benefits of subscription-based services and products which they would have to pay for in the absence of massive followers across social platforms. "Brands are becoming more interested in an influencer's objectives, the legitimacy of their followers and the worth of the content they publish" ("5 Benefits of Influencer Marketing," 2018). A decision to invest in influencer marketing could yield an increase in return than traditional methods of marketing, particularly when the following of the influencer is an element of the brand's audience. Customers are much more likely to purchase products suggested by influencers they have a relationship with, trust and have developed an emotional connection through experiencing their lives by regularly watching the influencer's social media posts.

Another major benefit of being an influential person is that have the opportunity to alter the lives of others by inspiring others, assist communities make a difference to our world's in a constructive direction. "People are drawn to influencers for advice, inspiration or to assist in guiding their style. The list of reasons goes on. It is crucial that you are aware and to be aware of the possible things you might make or say that could result in the loss of loyalty from your followers to decrease" ("5 Benefits of," 2018,).

To achieve a significant success as an emerging influencer, you must be successful based on your own merits and also the quality of your content. After you've made it as an influencer of your own, you'll be able to connect with an even larger audience in the event that you share knowledge with people around you, pass on new skills for others to aid others solve their problems. Furthermore, you'll be able to effect positive change in the world by being a well-known following since your noble intentions will be recognized by millions of people, and your followers will be financially a part of the worthy causes you promote.

Influencers are able to generate thousands of dollars to charities . They could play an important part in making the difference in people's lives through financial aid to help those who are struggling to overcome difficulties and challenges. People look to influencers to get inspiration and tend to share their wealth when they see their favorite influencers enhancing communities and enrich the lives of people in a way, for example, by giving money to charities that help the hungry.

Fourth, being an influencer offers tremendous opportunities for growth, learning and growth. Being an influential person, it is your job to be able to absorb and integrate practical knowledge and practical experience pertaining to your ability to be successful in the current climate of economics. This is due to the fact that you'll need to offer high-quality economic value to the people you want to reach to succeed being an influential. Growing your company's image to the next level requires you to understand the most effective methods for marketing to influencers and will require you to use your brain's capacity and your creativity to put out more appealing and captivating original content that offers unparalleled value to your targeted market.

Being an influencer can lead to a the satisfaction of a busy, fulfilling and rewarding life. If you are a successful influencer you'll be able to interact with other influencers obtain lucrative endorsement deals from companies, and discover many opportunities for growth of your brand. As you gain more recognition as an influencer who is successful you'll gain more followers, build more skills, accumulate more practical experience, generate additional revenue streams, and get more endorsement deals.

As an influencer you'll be able to make money from your interests and passions and transform what was once a costly endeavor into a full-time extremely lucrative, rewarding job that you built by yourself from nothing. Being an influencer will profoundly alter your life in all aspects particularly as you become more well-known as an expert in your chosen area, and exponentially increase your followers, influence many more people, and turn your dreams into reality.

As an influential influencer with a reputable name who has achieved a high level of fame and power, "you will have a platform that allows you to expand, learn and add subjects that are not covered in your content. The more diverse content you are able to offer people who are following you on social media, the better you can expand your influencers' social media networks" ("5 benefits of" 2018.). When you become an influential influencer online and achieving the highest level of popularity leverage, you not just will improve every aspect in your daily life but could increase your influence to help others to thrive, grow, and prosper. There are numerous advantages to being a successful influencer on the internet and having a huge leverage on fame will far surpass the time spent working as an employee, in every its aspects.

Earning a significant amount of money online in order in order to be able to positively enrich all aspects that you live, be in control of your professional life and even be an influencer is becoming more feasible than ever before. It's because it's now easier than ever before for influencers to make money through their websites and control their lives, especially since they are still in the initial stage of building their brand even though they may possess the privilege of producing brand-specific content on a regular basis before their brands start to gain significant momentum. This means that the potential influencer will never have to settle for in a gruelling, exhausting and unsustainable minimum wage job that doesn't even provide a basic salary to afford accommodation. The rising influencer can choose to make use of his income abilities to generate significant revenue even when the brand is during its first few months and not making any income. The rising influencer is typically an all-time content creator in the capacity of a creator, but has a high-income earning ability that can be leveraged to generate significant revenue when they are in the beginning phases of establishing his business.

In this age of digital technology today, it is easier than at any time in the past to acquire high-income capabilities even if you do not have the financial resources required to fund an education at a university. It is essential to acquire the highest income level in case you are planning to offer significant market value to your clients through the creation of a profitable niche brand. Acquiring high-income skills could be advantageous when you intend to build an established reputation and a substantial customer base to make a hefty income. A few of the many high-income abilities that don't require qualifications or degrees include high-ticket closing writing, copywriting, paid-speaking and coaching, as well as consulting programming, as well as digital marketing abilities.

## Conclusion

Like all entrepreneurs, Musk saw his fair number of difficulties. Inexperience in business led to Musk having to leave his initial venture because he was more of an engineer, rather than an executive. In the end Musk realized that in order to become successful in business it was necessary to be able to wear both caps.

Musk's advice about how to succeed is plentiful and varied, his most memorable advice would be to think outside the box. Musk loved reading books, and he knew many methods to solve the problem. Musk knew that failure was and will always be a possibility. He was worried about the outcome which wasn't in his favor, but it did not hinder the things he desired to accomplish. He just tried to reduce the fear to ensure that it didn't interfere with his thinking process.

There are a lot of instances where Musk didn't let minor factors stall his projects. As a millionaire, having private aircrafts or an island for his own use was nothing significant. If SpaceX requires more space for their rockets to fly, Musk has found an island that could create the space. He once flew his private plane to France to bring a machine which would accelerate production.

One of his most memorable creative moment of improvising was when he rented the truck that had fridge as the Tesla prototype had for testing in cold temperatures. It is said, "Where there's a will there's a means," and no one understands more than Musk.

Musk is known for dissing people, because he has his hands in many pie-cutter's. Here's a look at what Musk is currently working on, which has raised the competitive standard to a point where people are afraid to venture into any business run by this billionaire.